What They're Saying About
Time Based Security

"This book is really right!"

Dr. Fred Cohen, Principle Member Technical Staff,
Sandia National Laboratories; Inventor of Computer Viruses

*"Time Based Security is brilliant. Revolutionary thinking! Time
Based Security is to computer security as gunpowder was to
warfare. For the first time, those who would defend critical
infrastructures and priceless intellectual property have a manual for
defeating their attackers, and doing so in a cost-effective fashion.
The heart of this book is about the relationship between detection
time, sunk costs, and sufficient security--this is essential reading.*

Robert D. Steele, President, OSS Inc.

"Stimulating"

Dorothy Denning, Professor, Computer Science,
Georgetown University

*"Time Based Security presents a simple, common sense approach
that virtually anyone can use to apply to information assets."*

Lloyd F. Reese, CPP, CISSP, Program Manager

*"Mr. Schwartau offers an intriguing process to information systems
security which must be seriously considered when developing,
baselining, and/or testing the protection mechanisms of today's
systems. He explains why fortress mentality and the old ways of
security have not worked and provides an alternative, which is an
integration of new ideas and the tested ideas such as risk
management. His Time-Based Security Model can be nicely
integrated as the "other side of the coin" to compliment the
penetration testing in a more systematic and cost-effective process."*

Dr. Gerald L. Kovacich, CFE, CPP, CISSP, President,
Information Security Managem͏ ͏ciates

TIME

BASED

SECURITY

Winn Schwartau

interpact press

seminole, florida

Interpact Press, 1999

Practical and Provable Methods to Protect Enterprise And Infrastructure, Networks and Nation.

Schwartau, Winn

1. Computer Security, 2. Information Security 3. Network Security
4. Internet Security. 5. Hackers 6.Network Management 7. Information
Warfare, 8. Critical Infrastructures 9. National Security

Table of Contents

Recent Writing by Winn Schwartau

Books
- Information Warfare Volume 3: Technologies and Solutions (coming, 1999)
- Blitzkrieg.Com (coming 1999)
- Information Warfare: Chaos on the Electronic Superhighway (1994)
- Information Warfare: Revised (1995)
- Information Warfare: 2nd Edition (Thunders Mouth Press, 1997)
- The Complete Internet Business Toolkit (w/Chris Goggans, VNR 1995)
- Mehrwert Information (Schaffer/Poeschel, Germany 1996),
- Firewalls 101 (DPI Press 1996)
- Terminal Compromise (out of print, 1991)

Shorts
- The Toaster Rebellion of '08
- CyberChrist Meets Lady Luck
- CyberChrist Bites the Big Apple
- Almost 200 Articles (see www.infowar.com)

Book Contributions
- Introduction to Internet Security (DGI/MecklerMedia, 1994)
- Internet & Internetworking Security Handbook (Auerbach, 1995)
- Ethical Conundra of Information Warfare (AFCEA Press 1997)
- Something Other Than War (AFCEA Press, 1998).
- CyberWars: Espionage on the Internet (Plenum, 1997)
- Happy Hacker (American Eagle, 2nd Edition, 1998)
- National Security in the Information Age (Olin Foundation, 1996)

Acknowledgements

Time Based Security has been a long time in the making. Without question, I want to thank Bob Ayers, a close friend and superior security professional for introducing me to the concepts of PDR. We spent many long hours in Poland, Sweden, England, and all over the United States discussing and arguing what I subsequently named Time Based Security. Bob doesn't agree with everything in this book, and that is good. I hope he will keep pushing the envelope of debate for the industry.

I've been thinking about TBS for a long, long time, and every now and then it all becomes so crystal clear. (Aha! That's how it works!) Suddenly the world of security (physical or otherwise) is completely lucid. And then I wonder, if TBS is so insanely simple (as it is), then why has no one else thought of it before? Just as no one can explain why a particular inspiration occurs at a particular moment in time, I cannot explain Bob's PDR inspiration or the past failures of the human race; I prefer to recognize success.

You may also discover that I have unintentionally become something of a modern day Luddite. Technology for technology's sake is sheer insanity, and too many people pile the bits and bytes miles high in search of technical euphoria. I am annoyed by incessant voice mail, computerized telephone operators, PBX's without a <Back> button, Windows installations from Hell ... this could be a rant. You will find that I look to the simpler non-technical solutions first and prefer waiting to add technology. Fair warning.

Lastly, during the evolution of this book, I asked for comment by friends and colleagues throughout the industry. I thank them for their time and thoughtful responses. Among those who greatly assisted me are Dorothy Denning, Professor of Computer Science at Georgetown University, Dr. Gene "Spaf" Spafford, COAST, Purdue University, Dr. Mich Kabay, Director of

Education, International Computer Security Organization and Ed Glover, Sun Microsystems to name a few.

In the last twelve months, I have had the honor and privilege to present the concepts of Time Based Security to thousands of security professionals in the United States and Europe. I thank these wonderful audiences for their engaging and challenging feedback, which further helped the evolution of Time Based Security. Organizations such as Security Dynamics, Inc., BAI, Verisign, ISACA, ISSA, Computer Sentry, Inc. Internet Security Systems, Inc., and Sun Microsystems, have also been incredibly supportive. Thank you all.

For those of you who don't know her, Ms. Infowar, Betty O'Hearn has been an absolutely invaluable executive assistant who keeps me both in and out of trouble. ☺

Lastly, I want to thank my wife, Sherra, who after twenty-years of marriage still tolerates three laptop computers wired to the 100mb connections in the headboard. I love you.

Winn Schwartau

Forward

In 1993, I took over as the Director of the Department of Defense Information System Security Program, or DISSP, a Defense Department wide program established to improve the security of DoD's computing and telecommunications infrastructure. Within a very brief period of time, I became increasingly concerned at what the DISSP was discovering: that after many years of effort, and uncountable dollars invested, the security of the DoD's computing infrastructure was abysmal.

Let me explain how I arrived at this judgement. Two of the security services I established in the DISSP were designed to measure the security posture of DoD's operational computer infrastructure. One service was called The Vulnerability Analysis and Assistance Program, or VAAP; the other was called the Automated System Security Incident Support Team, or ASSIST.

The VAAP used publicly available attack tools downloaded from Internet sites to perform actual penetrations of operational computer systems. After two years of performing VAAPs, we had attacked well over 18,000 DoD computers. What we found was frightening:

- We were successful in gaining control of the target computer 88% of the time. (~15,840 success attacks)
- Our successful penetrations went undetected by DoD system administrators and system operators 96% of the time. (They only detected 633 attacks.)
- For the very few attacks that were detected, 95% of the time the system administrator or system operator took no action in response to the attack, i.e. they did not even report it. (Only 31 attacks created a reaction to them.)

When we analyzed the data collected as a result of over 18,000 penetration tests, we were able to demonstrate

iii

unambiguously that only one out of every 500 successful computer and network penetrations or break-ins was ever reported.

Working in concert with the VAAP was the ASSIST. The ASSIST was established to provide assistance free of charge to DoD personnel who were the victims of a computer attack. ASSIST personnel provided telephonic help, software, or, in critical cases, would deploy to DoD sites suffering from computer penetrations. In 1995, ASSIST received 555 reports of successful penetrations of DoD computers by unauthorized intruders.

Since the VAAP indicated that only one in 500 successful penetrations were reported, and the ASSIST program received 555 reports of successful penetrations in 1995, we were able to state with a high degree of confidence that the US DoD computing infrastructure had probably been successfully penetrated over 250,000 times in 1995 alone.

Something was horribly wrong. Why, after years of trying to build and operate secure computing systems, was the computer security posture of the DoD so bad? More importantly, what could be done to correct it?

The VAAP figures clearly highlighted the problems. Here is what they told us:

- Despite massive expenditures over some two decades, the US Department of Defense was unable to build and operate a system that could resist an attacker using free and publicly available attack tools. (VAAP Penetrated 88% of the systems attacked).
- The US DoD did not even know when its computer systems were under attack (successful penetrations went undetected 96% of the time).
- The US DoD did not know what actions to take in response to an attack (no action was taken in response to successful penetrations 95% of the time).

The underlying cause for the systemic security failures we so graphically demonstrated was an erroneous security philosophy that the US DoD had been pursuing since the requirement for computer security was first identified. That philosophy was 'Risk Avoidance'.

Since the philosophy of Risk Avoidance stated that a computer system should not be fielded if it contained any exploitable security flaws, the DoD had spent large amounts of time and money trying to build 'secure' or unpenetrable systems. The DoD confidently assumed that since they had made a large multi-year investment to secure their computer systems, it was axiomatic that they had been successful.

When the VAAP results were released in 1995, the initial reaction by senior DoD officials was disbelief; and disbelief was followed closely by denial. There was strong reluctance to accept the existence of a systemic security deficiency, especially one that would necessitate a massive investment, in fiscally austere times, to remedy.

The VAAP and ASSIST programs quantified and statistically proved that DoD's confidence in the security of its computers was unwarranted and that the many years of diligently subscribing to a risk avoidance philosophy had failed. Once this failure was clearly proven, we could address the reasons behind it

There were several unspoken assumptions on which the Risk Avoidance philosophy was based, or that were at least assumed to exist by the computer security community, if the long, laborious and expensive pre-deployment examination and testing of a computer system was to be justifiable. Unfortunately, these assumptions were wrong. The most critical of them were that:

- You could infer the security of a networked computer by testing it in a non-networked environment.
- You could infer the security of an operational computer by testing it without applications software running on it.
- You could make security judgements without knowing what was in the source code of the software running on the computer.
- The operational computing environment itself (systems and applications software, network connections, users, etc) was static.

In retrospect it seems quite inconceivable now that these assumptions were not seriously questioned. Unfortunately, one of the principal attributes of large government bureaucracies is that they are not usually known for critical self-assessment, and the DoD computer security community was no exception.

Before the untimely end of the DISSP - dismantled, as some believed, to spare DoD senior officials further embarrassment - a new security philosophy had emerged. Security was not merely a 'state', but a process that consisted of three fundamental components: protection, detection and reaction.

PDR represented a major philosophical shift in the theories of Computer Security. PDR was based on the assumption that it was impossible, either technically or fiscally, to build and operate a large DoD-wide 'secure' computing environment and that no security safeguards could resist a dedicated penetration attempt by an adversary who had an unlimited amount of time to attack the systems security defenses.

PDR postulated that the only true metric of the security of a system was the 'time' it took for a dedicated attacker to break the security mechanisms. The longer it took, the more 'secure' a system was.

Unfortunately, before the PDR methodology and the associated technical components under development to implement PDR within the DoD were fielded, the DISSP program was abruptly terminated and its human resources reallocated to other programs.

I am very pleased and honored that Winn Schwartau selected the PDR concept as a foundation on which to develop the much more mature 'Time Based Security' model, and flattered that he asked me to write this preface. Time Based Security provides quantifiable metrics that the computer security community has always lacked. Because of this, it will enable rational security investment decisions to be made for the first time. For example, which of the following statements has more value to the decision-maker:

- There are measurable, reasonable and provable mechanisms in place that can help stop an attacker before he has time to exploit a penetration (Time Based Security metrics), or

- The system is considered to be a 'B2'system (Orange Book Metrics) or,
- '...we can describe the extent to which the PP or TOE can be trusted to conform to the requirements' (New Common Criteria)?

The beauty of Time Based Security is that it is understandable to both the security practitioner and the decision-maker who must make investment decisions. Until now, computer security professionals have been unable to communicate security concepts in terms that decision-makers can understand. As a consequence, they have been viewed suspiciously as 20th century alchemists practicing an arcane art, understood by few and ignored by most. Time Based Security enables this much needed communication of concepts and quantification of terms to take place.

Bob Ayers

An Introduction to Time Based Security

You get up every mornin' from the 'larm clock's warning
Take the eight-fifteen into the city.
There's a whistle up above, people pushin, people shovin',
And the girls who try to look pretty.
And if your train's on time, you can get to work by nine.
And start your slavin' job to get your pay.[1]

There's never enough time. Is there?

We don't have enough time to spend with family. Not enough time to get to the airport. No time to read for pleasure. No time for the movies. Never enough time. That's life in the 90's.

"Make time." "Find the time." "What time is it?" "What time do I have to be there?" "I don't have enough time."

Phew. And that is how many of us find our jobs today. Filled with an apparent infinity of things to do in zero-time. Stress. Noise level. Priorities. Which action-item is worthy of my time? What can wait? I don't have enough time. And what about the kids? Or was Harry Chapin right?

"My son turned ten just the other day,
said thanks for the ball, dad, c'mon let's play.
Can you teach me to throw,
I said not today, I got a lot to do.
He said, that's OK.
And as he walked away, he smiled and said,
I'm gonna be like him. Y'know I'm gonna be like him." [2]

Networks. Computers. Faster computers. More bandwidth. Higher speed I/O. Nanoseconds. Faster computers yet. Picoseconds. Femtoseconds. Obsolescence; the time before your latest purchase is considered technically obsolete. Light speed. Quantum Computing. Moore's Law. My heavens! It was bad enough when

---

[1] *"Takin' Care of Business," Bachman-Turner-Overdrive, 1973*
[2] *"Cat's in the Cradle," Harry Chapin, 1974*

Moore's Law dictated that given the same dollars spent, we were doubling our computing power every eighteen months. That was fast. Change. Really fast change. And then in 1997, that high speed rate-of-change changed, effectively doubling itself.

Hyper-warp-speed Moore's Law now dictates that technology power doubling occurs every *nine* months. Too fast. Too much information. How do I choose? Not enough time. Not enough time... not enough time...

Yet we are all expected to get our jobs done on time. The report has to be to the boss on time. The client needs his presentations on time... or else! Get to work on time. And what about our spouse? Time for dinner? TV Dinner? Microwave the popcorn before bed 'cause there's not enough time to sleep before the next day's time crunch?

And then there is our own bastard stepchild: security. "Security can wait." "We don't have time for security now, we'll take care of it later... when we have more time." Right. Doesn't happen. What about the hackers? "Oh, don't worry; we've been safe for so long now, nothing will happen." Arrogance. "Security? Who needs it. It only costs us money and time." Apathy.

Time. It's all about time.

Time Based Security is not going to tell you that your organization needs security. You already know that. I am not going to proselytize about how bad the problems are. We have all written extensively on that subject for the last decade. You know what's going on. *Time Based Security* is an entirely different approach to security than you've ever seen before. I assume you know you have to do something to protect your information resources, your intellectual and/or electronic assets, and the integrity of your corporate infrastructure or the very fabric of national infrastructures themselves. There is no preaching from the bully pulpit here.

Time Based Security is your handbook for protecting intangible things of value that have no physical substance. The onslaught of the Information Age has created new techno-environmental conditions that are critical to understanding our jobs:

Modern society and most of our organizations have crossed a critical historical threshold. Our corporate and national vulnerability has entered an uncharted domain where our reliance upon the technology, the networks, communications systems and infrastructural interdependence has exceeded our ability to live without them.

Current macro-economic philosophies do not provide a mechanism by which to determine both the ever-changing entropic and anti-entropic value of information. By and large information evaluation is not a measurable line-item to be found on corporate balance sheets.

Time Based Security is a non-technical examination of the very foundation of the technical realities of the networked society. It is designed for a wide audience with varying skill sets, backgrounds and business needs.

You do not need a masters degree in anything to understand the concepts presented in *Time Based Security*.

You do not need to be a sophisticated network engineer to immediately apply the premises of Time Based Security. In fact, within minutes of reading this book, you will be able to accurately gauge just how secure your organization is – or isn't – and know what steps you need to take to improve the situation.

Time Based Security is meant to be read by computer and network manufacturers, managers, company owners, politicians, policy makers, network administrators as well as security professionals. The value is in the ideas and concepts, which work, in countless different environments.

The premises of *Time Based Security* apply to small networks, large networks, unimaginably huge networks, ISPs, heterogeneous corporate Enterprise Networks as well as the independent and interdependent critical infrastructures, which glue our society together.

Time Based Security underscores the fact that the private sector, the military, and intelligence organizations in every developed and emerging nation-state are no longer at odds. We are all in this together. We share infrastructures. We share connectivity. *Time Based Security* provides general models to assist in protecting the unrestrained technical globalizing of the planet.

Time Based Security is written in "bite-sized" chunks, to be digested one thought at a time. You do not have to sit and read *Time Based Security* all the way through in one sitting. In fact, I recommend that you spend time in between each small "Chaplet" to think about the ideas posed and how they can apply to you in your specific environment.

Also, don't miss out on the opportunity to "Think Big!" When you apply the concepts of Time Based Security to your specific environment, think about how they also might apply to such worlds as entire Traffic Lights Systems, Power Distribution, Cellular Communications, Transportation Networks and other critical infrastructure components.

Time Based Security is meant to be a tool, but I do not claim that the concepts here are perfect tools. They are brand new; just recently oozed out of stressed-out cranial matter. They represent a fundamental paradigm shift in the way that we perceive security and its real-world application. I hope that you will think about security in a new and different way than you have in the past and use some of the Time Based Security concepts to defend your systems.

Time Based Security is not meant to replace everything we have all learned about security. It is meant to be an adjunct to existing approaches to security, and then it is up to you to independently choose what steps and actions are appropriate.

Hopefully *Time Based Security* will provide a foundation by which to further evolve the state of the art in years to come. I am grateful for the opportunity to present these ideas to you, and I welcome your feedback, which I hope to include in a future updated version of this book.

Winn Schwartau
November, 1998

Time Based Security

1
Profile of a Security Model:
Promise and Hope of
Time Based Security

Computer security has finally, at long last, become mainstream. When ads on CNN use espionage and hackers to justify the need for network and computer security, and security companies have Wall Street evaluations in excess of a billion dollars, security products now stand on nearly equal footing with an Intel CPU, Microsoft operating system or cellular phone. All 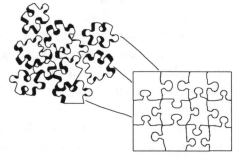 that we have to do is make these others secure. What may be in question, though, is how well current products and services really provide protection for untold trillions of dollars in intellectual property value. This statement is not meant to say that any manufacturer of security products is derelict, or that they produce bad products. It is meant to make us think about the overall approach we give to information and network security.

As professional security practitioners, we need answers. We need tools to do our jobs. We need models to achieve our goals. We need these tools and models to protect our corporate assets, networks and enterprises. We require strong models and products and services to protect the interdependent critical infrastructures: transportation, finance, communications and power. The military, intelligence community and governments (federal, state, and local) also require answers, tools, and modern approaches to adequately protect their resources; for national security, public services, and day-to-day operations.

When all is said and done, we find that current methods and models upon which security is approached have gaping holes which do not provide adequate levels of measurable protection. The Time Based Security Model (TBS) hopes to correct this problem by providing:

- A process methodology by which a security practitioner can quantifiably test and measure the effectiveness of security in enterprise and inter-enterprise environments.
- A quantifiable framework so that the security professional and management can make informed decisions as to where to smartly invest their security budget dollars.

Information Security is a young discipline - only about one generation in human terms, but about twenty or so generations in technological terms according to Moore's Law. Throughout its evolution, computer and information security has been largely based upon a single postulate, which is now suspect in its efficacy: fortress mentality. We have historically dedicated our security programs to build Fortress-like systems to *'keep the bad guys out,'* rather than encourage bi-directional commerce. Modern business communications and commerce are symmetrical in nature; bi-directional information flow is a fundamental requirement for modern enterprises today.

Many professionals now understand that the concept of Risk Avoidance is an untenable approach to true systems security. At the national infrastructural level, too, inter-enterprise connectivity is bi-directional for both information flow and control signals, and thus also falling victim to the failings of fortress mentality.

The findings of many works from 1990-1997, "Computers At Risk"[3], "Information Warfare: Chaos on the Electronic Superhighway,"[4] "Information Warfare: 2nd Edition,"[5] "Defense Science Board Report,"[6] and the PCCIP Report,[7] (among others) come to the same conclusion on a larger scale: the infrastructure of

---

[3] *National Resource Council, 1990*
[4] *Winn Schwartau, Thunders Mouth Press, 1994*
[5] *Winn Schwartau, Thunders Mouth Press, 1997*
[6] *Defense Science Board, 1996*
[7] *President's Commission on Critical Infrastructure Protection, 1997*

the United States is a critical national asset, it is very complex, highly interconnected and vulnerable to various intensities of assault.

Most recently, the President's Commission on Critical Infrastructure Protection (PCCIP) underscored the fact we need new solutions to solve the security dilemma, and none are on the immediate horizon. The TBS Model suggests otherwise.

The security approach defined by Time Based Security offers practical architectural solutions such as that described in *Surviving Denial of Service* in a later 'chaplet' in this book. The mechanism offered for defending against and surviving such DoS attacks is but one of a class of solutions needed to effectively deal with the real-world infrastructure security concerns of a networked society.[8] Any security model will bring with it several benefits, and Time Based Security offers a host:

- *Simplicity*. Time-Based Security is conceptually simple and offers extreme utility to vendor, consultant, integrator and customer. You can immediately apply and use TBS, regardless of your technical skills. (Of course, the more you know, the more the model will achieve.)
- *Utility*. Time Based Security does not hinder network operations or the ability of administration, management and users to do their job. Other security models and approaches can negatively affect network efficiency and user productivity.
- *Scalable*. Time Based Security offers protection at levels from the smallest network, to the largest of critical infrastructures.
- *Measurable*. Time-Based Security is a logical and comprehensive methodology to measure and quantify the effectiveness of security and support security budget decision making.
- *Quantifiable*. The model's quantifiable metrics provide replicable mathematical tools to measure the integrity of solutions to system security problems.

---

[8] *"Solving Denial of Service on the Internet", by Winn Schwartau, 1997. First published in the Proceedings of the 20th National Information Systems Security Conference, (NIST/NCSC), October 8, 1997.*

- *Provable*. Time-Based Security uses simple, basic mathematics at its core, which are replicable in disparate environments.
- *Supports Management*. Time Based Security offers mechanisms to allow management to make informed budgetary decisions on information resource and systems defensive spending and risk.
- *Adds Value*: Of course, the security model must offer something new to the practitioner, and something of value over or different from current approaches. Based upon the reactions of thousands of people so far, Time Based Security does just that.

It is my hope that Time Based Security will do justice to Bob's original concept and my enhancements will make Time Based Security workable across the spectrum of the information security community, at the enterprise and at the infrastructural level.

But first, we have to understand the changing nature of business today to appreciate why current security models and approaches are not adequate to meet the complex challenges they face.

2
Business Cycles, Time and Security

Capitalistic business practices are all about time. Consider:

Inventory: Businesses want to maintain as small an inventory as they can. GM and other industrial-age manufacturers measure their new unsold cars in time. "We have 41 ½ days of inventory." There is a specific dollar value assigned to that time, as well. Food retailers measure their inventory very carefully, using time as a metric, especially perishables such as milk which contains a can't-be-sold-after date. Medicines carry dates when they can no longer be trusted. Any miscalculations create waste and therefore lost profits.

Just-in-Time Manufacturing. Manufacturers like to carefully plan and optimize their incoming parts supplies so they don't sit around on shelves for eternity. Using MRP and other automated systems, purchasing agents today prefer to have supplies come in every day in smaller quantities than in huge month-long or year-long shipments. It saves money. The object is to get the parts into the factory, build them into the value-added product and ship the product out the doors as fast as humanly possible. Profits go up when the time between receiving and shipping goes down. That is a simple time-based reality.

And then there's Wall Street. American investors measure profitability of companies in small chunks of time – quarters of a year. Unlike some of our more visionary global friends and business competitors who plan and project industrial and national growth and wealth over decades, we seek out short-term profits and immediate gratification.

In the on-line world, we *also* expect immediate gratification. We expect and demand that Web pages launch instantly. "It takes too long to load." On-line commerce is absolutely about time. If an on-line payment cannot be made and authorized in a few seconds, many of us will walk away frustrated. "It must be broken." That's why commerce servers use cryptographic accelerators. It takes the commerce server a small, but finite amount of time to cryptographically decode an incoming payment message; on the order of 100 to 200 milliseconds. If just one person is making the request, he is processed very quickly. However, if the server is hit with a large quantity of commerce or payment requests all at once, a queue is created, and someone is going to be waiting a long time. A long, long time. So, cryptographic accelerators are hardware optimized to handle the complex mathematics of public key encryption systems and speed up the process by as much as 100 times or more.

War is also about time.

During the Gulf War, we measured the effectiveness of the land campaign against Iraq in hours; the so-called 100 hour war. History tells us about the 100 year war, too.

The Cold War chillingly told us that the planet was a mere 18 minutes away from thermonuclear extinction. That's what the Cuban Missile Crisis was all about – time. It gave our adversaries a distinct time-advantage for nuclear missiles to reach the US mainland, and that was unacceptable. How far was a day's march to meet the enemy in Napoleonic or Civil War days? How many days or weeks of supplies are on hand to keep the troops in fighting shape?

In high-speed wars, the generals want and expect real-time intelligence, use powerful technology to evaluate the data so they can make decisions, and transmit orders to the front-line combatants over very fast and very secure communications lines. Further, with the advent of the Digital Battlefield, it has been jokingly said that soldiers will be so wired with technology and smart weapons, they will require a Master of Computer Science before they are allowed to shoot.

In fact, when we think about it, most human endeavors can be effectively measured in time. Think about that for a second. (I am sure there are exceptions, as there are to any generality, but time is absolutely critical for most anything.) Our personal value is measured in time; dollars per hour; salary per month or year. Rates of investment, annual percentage rate. Time is the constant element throughout. And Cyber-wise, it's about bandwidth, just another way of measuring time: bits per second is the single biggest driving force behind the Global Ne twork. Speed. Time.

Now, consider the business cycle. (This also applies to war, by the way.) An entrepreneur (or established company) thinks it wants to get into a new business or product arena. It researches the field, analyzes the market, and indulges in competitive intelligence. (War: The military and intelligence communities must keep tabs on everyone, everywhere, at all times, in order to advise political and military leaders. They use technology such as SIGINT, Signals Intelligence and HUMINT, spies or Human Intelligence, and OSINT, Open Sources Intelligence, unclassified information available to anyone for the asking.)[9]

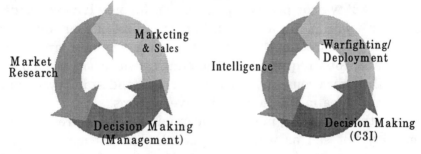

The top management of the company uses the market research data and finally determines it wants to begin selling Green Furry Things to a population hungry for them. They tool up the factories and begin making Green Furry Things by the millions. (War: The Pentagon, or Ministry of Defence in the UK, or whatever other military force is around, designs actions and responses to potential scenarios based upon the intelligence provided. In some cases, the military may be called upon to plan and execute an operation; peace-keeping, humanitarian missions, joint operations or maybe actual combat.)

---

[9] *Military intelligence is time-based as well. If a satelite photo shows ships in a harbor, five days travel away from a potential target, the window of protection is five days. If there are no new photos taken for three days, then the window of protection is only two days. Newer information is needed to rejudge the window or protection. Therefore, constant updated intelligence is a form of derfense and protetion – measured in time.*

The sales and marketing troops are released. The advertising and public relations begins an onslaught first announcing the availability of Green Furry Things at stores everywhere, and then convincing the public that they will be less than American if they don't have several Green Furry Things of their own at home, in the car and lavishly hung around their necks. The sales force will convince stores to stock up on Green Furry Things so they can fill the impending demand. (War: The troops are sent to the field; Somalia, Haiti, Panama, the mid-East. They keep the peace or shoot the enemy or bomb the bejeezus out of the desert for forty days and forty nights. They begin their mission.)

The circle, which represents both the commercial world and the business of war or peacekeeping, is continuous. It never ends. Thus, in both worlds, we are in an iterative state of constant flux; the conditions of free-market realities and geo-politics are ever-changing and thus require constant adjustment. Such is modern living on the precarious ledge between chaos and stability – that great unknown realm of complex adaptive systems.

For the next cycle, in the commercial world, the Green Furry Things may not sell as well as expected, thus (1) more market research and customer feedback is needed so that (2) management can make updated decisions and (3) the sales and marketing people can broadcast the new message... and so it goes on and on...over and over again. It never stops.

Or perhaps, Green Furry Things sell just fine, meeting or exceeding expectations, so (1) market research is conducted to see if the buying public wants Blue and Brown Furry Things as well, or perhaps Green Furry Things that belch on alternate Tuesdays. Whatever. (2) Decisions are made and (3) the sales and marketing troops again pounce upon the public.

In the military world, once a mission has begun, (1) constant and updated intelligence is required, along with instantaneous real-time battlefield feedback from the technology-armored soldier which is sent back to (2) headquarters where additional decisions are made and commands are issued. Then, (3) the deployed troops are issued updated orders. And so this cycle goes on and on...until the mission is completed at the local level ... or forever on the macro-scale of global geo-politics and national security.

Unfortunately, management sees information security as an unmeasurable bottom-line drain on profits, or an "insurance policy" against which actuarials are slim and hard numbers are more folk-lore than statistically defensible. Or, management sees security as an unnecessary evil or burden that interferes with getting the job done. Too many security professionals and security product vendors view security as a technical problem, thereby demanding a technical solution. Ouch. Consider two things:

What two words are used to hire network adminstrators and security people more than any other? "Hey, You!" In too many cases, the person with admitted word processing skills is seen as the obvious candidate to provide technical guidance. Being caught by the water cooler is the easiest way to be singled out.

Most organizations which have information security problems are not going to solve them by adding more technology as a first step. The additional complexity of sophisticated security products is likely only to confound ill-trained admininstrators, create a culture of mis-configuration and result in more problems than they began with.

Security, in reality, is about protecting the business process. The circle. Our jobs are to maintain the even flow of that iterative process which creates revenues, profits, and jobs. It doesn't matter in the least what product or service you are selling. It doesn't matter what industry you are in. All infrastructures are essentially the same and it doesn't matter what war you are fighting or humanitarian mission you be conducting. Security is only about letting an organization do its job efficiently or conduct its mission – all without hindrance.

Our security-oriented jobs are to keep that iterative process going uninterupted. Our jobs are to insure that no breaks in that circle occur – by any means. Imagine trying to spin a hula-hoop with an 24-inch gap in it. You will not be very successful. Our jobs are not to unnecessarily inbue organizations with technology. Rather, our jobs are to help understand the business process well enough, and coordinate with management so that whatever security efforts are put into place, (policy, process, technology), that the real goals of the business are maintained with as much purity as possible.

Time Based Security becomes an assessment tool so that the security folks can accurately and quantitatively report to management, so informed budgetary decisions can be made. For security, this is a first. Time Based Security offers guidance and direction to achieve these goals, partially through the use of technology, but only once many aspect of the business process are more clearly understood. You will see that TBS requires an understanding of the relationship between very specific organizational and informational assets to tune its effectiveness. This common sense perception has somehow been lost in the maelstrom of protective technology.

You will be able to immediately use TBS and determine how secure your organization is – without hours of expensive seminars and training, but it *will* take extra efforts to maximize its effectiveness. Next, though, we need to see how drastically the demands upon security professionals and technology have changed in the last few years.

3
Symmetric Security

I look back at the good ol' days when information security was so darned easy. Well, maybe not darned easy, but one whole heck of a lot simpler. Really, it was. And there are whole lots of security professionals and IT managers who also recall how security used to be so simple. (Or simpler...)

Trusted workers sat at text-only dumb terminals which were hard wired along a well defined single path to a mainframe computer where the data resided. No smart networks deciding by themselves what portion of each message should take which path to the final destination. Just a single wire. There were no floppy drives driving a stake through your other infosec efforts.[10]

At the mainframe, database or midrange server, a reasonably simple access control mechanism kept clerks out of the files where they weren't wanted. It kept salespeople away from payroll. The rules were pretty simple. You either had or didn't have access to certain resources and information. The central security management tool resided at the data base, was remotely managed and we tended to look at the world in terms of 'groups' by location, function or activity.

Ah, those were the good ol' days of "one-way streets" and uni-directional security. Users at the amber or green teminals only had one place to go, and that was that. No one was trying to specifically speak with a distant terminal. It had nothing to say. Sort of like a radio station which broadcasts (transmits)

---

[10] *Consider the following argument, then try it on your boss. You run or work in a networked company, right?* Yes. *And you use print servers and routers and distributed applications, eh?* Yes. *And you don't want viruses to infect your networks, do you?* No, of course not. *Do you want employees bring in games to work?* Nooo.... *And you don't want employees taking home sensitive files or privacy oriented materials without permission, do you?* No. *Then, why do all of your machines have floppy drives?*

Programs should be centrally managed for copyright adherence, revision control & management, and operational needs. Instead of automatically creating a security hole by proliferating floppy drives, make "none" the standard, and only give access to floppies to those who really need it. Maintenance staff could use parallel port or other portable I/O devices when necessary.

information-rich radio-waves into the aether for anyone to "listen in on." The radio receiver in our house, car or blaring into nauseatingly colorful headphones, has nothing to say either.

But today? Everyone wants to talk to everyone. Everyone wants to publish and be heard. (That's a technical comment, not a social one...) That's' what PC is: independent processing from a central computer. Distributed application. Technical autonomy on every desktop. And than meant managing information flows in more than one direction. The PC was both a source and a destination as networks proliferated in the late 1980's. With the Internet, and with business-business applications on a single campus, or virtual networks, the matrix of communications paths multiplies, and we now have to manage security symmetrically. Security complexity in the network increases dramatically as the number of possible communications paths spiral into a web of cotton candy density. The architecture blurs from physical to logical to virtual.

And finally, perhaps more than any other single application, email redefined security needs. Symmetry What can be done at one node of a network must be able to be done on the other nodes. If I can email you and you cannot email me back, that's using a computer network as a pager. (Dumb.) Two way pagers – email; same thing. Symmetry. Architectural symmetry of function, and from a security standpoint, now I have to manage it in two directions. From A>B and from B>A.

So, the security-control needs by many organizations are terrifically complex and we find that instead of a simple one way series of groups to manage, we have to deal with  dozens of different real-world operating environments and constantly changing conditions:

- Internal staff with access to physically local logical resources
- Internal staff with access to physically remote logical resources across dedicated internal networks
- Internal staff with access to physically remote logical resources across external public switched networks

- Internal staff access control to remote logical resources across external public switched networks
- Internal staff bringing remote data and information into the internal networks
- Internal staff publishing to remote logical resources
- Internal staff removing internal electronic resources and sending them to remote locations
- External staff requiring access to some, but not all, internal network resources across public switched networks
- External staff accessing remote resources
- Internal and external staff access control to remote logical resources across external public switched networks
- External traveling staff requiring access to some, but not all, internal network resources
- Business partners requiring access to some, but not all, internal network resources.
- Customers requiring access to some, but not all, internal network resources
- Customers requiring access to all public company publications, but not any internal resources
- Potential customers requiring access to all public company publications, but not any internal resources.

And these are only a few of the possibilities! Try to list the number of security paths and functions in your networks, then make sure you look at it from the other direction, too, to insure you're tackling the symmetrical nature of networking in your analysis. Keep in mind that commerce is symmetric in nature. Merely connecting to a website invokes security questions that must be answered symmetrically.

A: *On one hand, can I travel to a specific website? (Or is there a set of URL blocks at the firewall; or banned sex-sites using contents filtering.)*
B: *Can I receive email from those same domains? And will I accept applets from every URL?*

Or...

A: *May I send out executables and small programs to other people I know? (filter outgoing portal; contents analysis)*
B: *Can I upload programs into my desktop from those same people?*

And most importantly, when buying something over the Internet, the cryptographic nature of strong user authentication or session keys (etc.) is a successful two-way dialog between the two machines, resulting in a 'trusted' [11] relationship

Many organizations find their own access control needs even more complex, but the point is real world security implementation must be symmetrical in nature. Security must be implemented in two directions for each electronic nexus: the information going both in and out of the organization, the people going both in and out of each nexus; the myriad combinations thereof.

So, how does Fortress Mentality work to reflect the business needs of symmetric processing? To get the answer, we will need to look in at a voyage that began more than 3,000 years ago.

---

11 Trusted means something very specific to some people. It implies that the software or security product has gone through specific testing, and been proven to some extent to be secure. This often refers to operating systems and specific applications such as data bases, and a specific level of trust has been established by the US Government's testing authority, the NSA. Trusted relationships in networking are often established for short periods of time, often for a second or less as with the transmission of passwords, financial transaction information. Longer duration trusts are needed for VPNs. These products are untested by any certifying authority (by and large) and are judged in the melee of the technical free-market economy.

4
Fortress Mentality: Walled Cities and Other Battles Lost

In 1200BC, Paris abducted Helen, wife of Menelaus, a leader in Greek Peloponesia. Under Agamemnon, the Greeks sailed to Troy and besieged the city for nine years in an effort to retrieve Helen. (No matter what she wanted out of the deal.) Troy was a walled city, capable of sustaining itself without outside assistance.

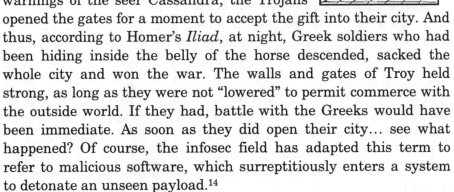

Seemingly bored with waiting outside the closed gates of Troy for almost a decade, the Greeks up and left. However, they left behind what the Trojans considered a present, in honor of the their 'victory' in the war: a great wooden horse. Despite the warnings of the seer Cassandra, the Trojans opened the gates for a moment to accept the gift into their city. And thus, according to Homer's *Iliad*, at night, Greek soldiers who had been hiding inside the belly of the horse descended, sacked the whole city and won the war. The walls and gates of Troy held strong, as long as they were not "lowered" to permit commerce with the outside world. If they had, battle with the Greeks would have been immediate. As soon as they did open their city... see what happened? Of course, the infosec field has adapted this term to refer to malicious software, which surreptitiously enters a system to detonate an unseen payload.[14]

A millenium later, 300,000 criminals, military conscripts and slaves commenced fortifying a wall 2,400 kilometers long; the Great Wall of China. What tourists climb upon today was completed in the Ming Dynasty, from 1368 - 1644. Despite being an

---

[14] *This is a perfect historical case where contents analysis would have solved a tremendous problem for the Trojans.*

average of almost eight meters tall, the Great Wall proved to be of little military defensive value against the marauding northern

invaders. Even today, we recognize how unsecure and permeable the Great Wall of China really is. After all, most of us have seen David Copperfield float right through it!

New technology was invented to make these defensive walls less effective than they were designed to be. Catapults, ladders and battering rams proved to be very effective tools. In the Middle Ages, castles were defended by wide moats teeming with hungry carnivores and high, thick bastions. While the defenders poured boiling oil onto attacking troops, catapults lurched massive stones at the fortifications in hope of breaching them. The problem the castles also faced was that they had to let down the drawbridge periodically to permit trade to occur.

And so it has gone throughout military history. Fortress Mentality. Keep the bad guys out. Build walls so high, they can't get in. Build them so thick, they cannot be battered. Build them so long they cannot be run around. But we are learning such approaches do not work. The French attempted to build a defensive line on the German border. Under the leadership of French politician and minister of war Andre Maginot, (1922-1924 and 1929-1932) construction began in 1925. Thought to be impregnable, the Maginot Line was bypassed and later captured by the Germans in 1940.

Fortress Mentality.

It doesn't work. It is static. It cost millions of lives during the trench warfare of Word War I. How do you keep the bad guys out and let the good guys in? How can you make that differentiation reliably, regularly and efficiently? This same question haunts security and network administrators today. How can I build a static defense, a fortress mentality, that is manageable and will protect my resources, yet let the good guys in.

The Cold War witnessed more Fortress Mentality with the Berlin Wall. Perhaps more symbolic than militarily significant, nonetheless, people dug under it, crawled around it and flew over it in hot air balloons. And finally, like all of the others, it came tumbling down. So, when the ideas of computer security first came up in that era, Fortress Mentality was the leading candidate.

Where We Used to Keep the Information

Let's step back to the 1970's. Information Security was partially based upon an older physical model. The protection of sensitive hard copy information entailed placing armed guards in front of a locked room containing a padlocked file cabinet. Info-guards of the Cold War. Only the good guys, who were already inside the building could get inside the locked room and then unlock the secret filing cabinet. So, why not apply the same logic to information security? While the military has an endless supply of teenagers carrying machine guns, the private sector cannot take the same Draconian approach.

Computer security was initially a military concern, addressing saving US secrets from the Evil Empire. Espionage was about military secrets, weapons, war and national security. The penalty was death. Therefore, it's understandable that the early days of infosec echoed the traditional military views, attitudes and approaches. So, despite thousands of years of failures, Fortress Mentality was again employed to protect the intellectual asset valuables of America.

And that has been the approach we as an industry have taken to information protection and security. The history of security models and security standards ever so exemplify Fortress Mentality and Risk Avoidance. Information (intellectual assets, resources, and intangibles) is no longer protected with the vigilance of an army, and the actual logical location of the resources has also changed significantly. Information assets are not secreted away behind fences and gates, down hallways filled with nerve gas, or isolated in bomb shelters such as shown in the fanciful *Mission Impossible* movie.

Today, information is at the very periphery of our organizations. Networks are porous conductors of people reaching out to touch someone or something, a virtual inch or two away. We want the information to be accessible. We want window-shopping surfers to "Come on down!" and read our words of wisdom or make a purchase. We want visitors. But, we want nice, well behaved visitors.

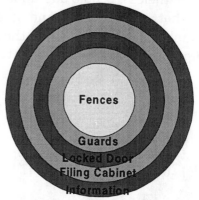

Where We Keep the Information Today

Because the networks are all tied together; because we demand instantaneity in our information lives as well as our real lives, the physical Fortress Mentality barriers of yesterday are gone. We don't have to pass by the secretary or the guards or a locked door to get to the Information Treasure Chest. It is sitting no farther away from us than the advertising materials most companies beg us to memorize.

And because we function in a symmetrical world of communications, the Fortress Mentality models and methods we have employed to protect our computers to date …just won't work. In fact, they can't work. Not alone, that's for sure.

5
A Brief History of
Security Models

Security models and methodology are all too often based upon premises that do not and cannot generate adequate solution suites to the problems faced in today's heterogeneous, symmetric networked world. We operate with so many different hardware platforms, operating systems, versions of O/S's, protocols and media, (heterogeneous) and symmetrical communications as described earlier, that no matter how hard we try, establishing a working benchmark for security has been elusive at best; a compromise at worst.

Fortress Mentality Security Criteria

- TCSEC (Draft) – 1983 (US Trusted Computer Security Evaluation Criteria)
- TCSEC (Final) – 1985
- TNI – 1987 (US Trusted Network Interpretation)
- C2++ 1989 (Proposed)
- ITSEC – 1990 (UK, Germany, Netherlands, Belgium)
- JSEC –1991 (Japan)
- MSFR – 1991 (US Minimum Security Functional Requirements)
- FC –1993 (US Federal Criteria)
- CC – 1996 (Common Criteria, US and Europe)

The TCSEC

It all began with a military funding effort from the Department of Defense and some very smart people who were hired to think about the issues of computer security. The outcome was the TCSEC (Trusted Computer Security Evaluation Criteria), a/k/a, the Orange Book[15], the first formal information security methodology. First debuted in 1983, the 1985 version became the guideline by which we were to secure America's computers; to secure the commercial and government sectors from bad guys breaking into their computers. And the way we told everyone to proceed was with a classification system from a fairly insecure 'D' level to the top, best of breed, 'A- 1' which included mathematical proofs and years and years of analysis.

TCSEC employs the concept of a Reference Monitor to provide information protection, such as implemented by Bell & LaPadula. In this model, system requests (process, files, etc.) are to be mediated by a so-called traffic cop - the Reference Monitor – before being executed. Whether at the application layer, with the Operating System or even at lower network and hardware levels, the Reference Monitor creates a range of problems for the both the security engineer and the systems user.

The Reference Monitor

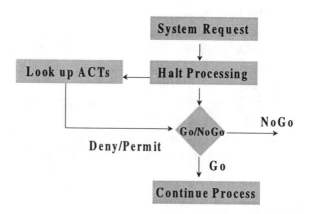

The Reference Monitor is a piece of software (or hardware, too) which listens to the system. When the system requests a file, or a process be invoked, the Reference Monitor halts the process or

---

[15] *The Trusted Computer Security Evaluation Criteria was first offered in a 1983 document, then formalized in the 'Orange Book' in 1985.*

request and queries an Access Control Table. If the ACT says that the particular user is permitted to conduct such activity, at such and such a time, on such and such as date, then the system proceeds, until the next request is made. If the permission is denied, the user is so told, and he cannot proceed.

In standalone computer systems, which the TCSEC addressed, this premise works. While the Reference Monitor model slows down the system by requiring a process to be halted until the mediation is complete, such a slowdown can be minimal (less than 10% system performance degradation) if only a few processes and users are to be monitored.

The TCSEC also calls for the generation of an audit trail of system and process executions, and the results of mediation. Granularity refers to the detail provided by the audit trail, and too much granularity in access control and permission rules creates additional CPU overhead and immense audit trails, thus further using up valuable system resources and time.

But even more so than the limitations of the Reference Monitor and the burden of extensive audit repositories are the inherent philosophical problems that the Orange Book presents to real world security practitioners.

TCSEC was aimed at standalone computer systems, (mini-computers and mainframes) before the proliferation of desktop and laptop computers. In addition, certain degrees of physical security were assumed to be protecting the computing resources – such as fences, barbed wire, dogs, locked doors, and teenagers carrying M-16's. Largely, TCSEC was for government and military use.[16] Despite encouragement from the government, the private sector never accepted their approach to security, and in many ways accounts for the lack of security implementation by the public since 1983.

In 1987, the National Computer Security Center published the Trusted Network Interpretation, supposedly an approach to apply Orange Book principles in a networked environment. This

---

[16] *That is why even though Microsoft NT claims to meet C2 TCSEC security, most professionals agree that NT is not secure in any meaningful way. Further, the NT C2 is only for standalone systems, not for networked environments.*

effort was even more of a commercial failure as it offered few real solutions.[17] After two years of study, the paper essentially came to the conclusion that "we have no earthly idea on how to secure a network."

Secured Operating Systems were the thrust of many TCSEC efforts. The National Security Agency accreditation of a secure Operating System is an exhaustive process that can take two or more years to complete. In the commercial world, this represents two or more generations of technology. (Moore's Law) Secure Operating Systems (for standalone environments) were often so restrictive in their operation that needed functionality was stripped away. In the private sector this was anathema.

Secure Operating Systems often required applications to be specially written for that Secure O/S, according to a revised set of rules. This would necessarily double the efforts of a software developer.[18] Consider CHOTS, the largest 'secure' network in the world, handling over 50,000 terminal users throughout the British Ministry of Defence. It is secure, to the rough equivalent of the Orange Book's B2 (multi-level labeling). CHOTS cost in excess of $1 Billion to build - but according to senior level military officials who are forced to endure it, CHOTS and its ten-year-old applications are useless. Thus, the circa-1987 DOS-like program applications relegate CHOTS to being the most expensive email system in the world.

Development of applications, which meet TCSEC specifications, can take just as long, an eternity in a world where Moore's Law reigns supreme.[19] Obsolete, often useless applications become the rule as was experienced in CHOTS, which was designed to meet an amalgam of the U.S. TCSEC specifications and the European ITSEC.[20]

---

[17] *In 1990, we designed COMPSEC II for Novell Netware, which subsequently was sold, evolved into Assure and is now the C2 component of Netware.*

[18] *This is the stated reason by Microsoft as to why they don't want to build secure operating systems for the commercial market or for export.*

[19] *Moore's Law says that given constant dollars, the power of computing doubles every 18 months. Recent developments by IBM and Intelhavs suggested that this period might be reduced to less than a year.*

[20] *The Information Technology Security Evaluation Criteria was a joint effort by the U.K., Germany, the Netherlands and Belgium.*

The application of Fortress Mentality to networks certainly has some limitations. Imagine, for example, attempting to use Reference Monitor mediation at a network nexus where symmetric communications are forever creating requests, opening files and launching processes. Bandwidth management is tough enough without imposing the overhead of a reference monitor. The system would come to a grinding halt. In modern software applications, (Object Oriented Programming) a program invokes perhaps dozens and more of associated files to function. Again, using highly granular Reference Monitor techniques, the performance would be unacceptably snail-like.

While TSCEC (and other criteria) do not have real world applicability in the commercial sector today (with some exceptions), we must acknowledge their contribution to the thought and evolution of computer and information security. Other similar security approaches were attempted over the years, but all were ultimately found to be similarly lacking.[21]

The resource, time and money expenditures required in Reference Monitor based security and application deployment is clearly unacceptable. The private sector is demanding both security and functionality, with security taking a back seat if efficiency or productivity are the price. Such is the history of computer security, and perhaps one of the reasons that we find ourselves in the late 1990's without a reasonable security model to apply or a metric to benchmark.

The Internet, Intranets, email and webification changed the rules of security virtually overnight and we hadn't done it right at that point either. For all practical purposes, our approach to computer and network security has been wrong. It may be determined in many cases that the best way to secure the electronic assets of an organization will be to eliminate all forms of protection. Counter-intuitive, yes. Workable? Maybe. We will see.

---

[21] *C2++ by Ken Cutler, MSFR by NIST, FC by NCSC, JSEC, The Common Criteria, etc.*

6
Lessons Learned

In the U.S. alone, over a period of twenty-plus years, we've collectively spent several billion dollars thinking about computer security. The government has made significant investments in both research and development as well as products, Many of those products have been total disasters.

LEAD, Low-cost Encryption Authentication Device was anything but low cost. For a mere $400 or more, this government sponsored product permitted a user to plug in a semi-smart-card to authenticate himself to the computer. Then there was Fortezza, based upon the classified Skipjack algorithm – part of the ill-fated and universally hated Clipper Chip efforts. In mid 1998, the government declassified Skipjack and said goodbye to their multi-billion dollar attempt at setting up their own Defense Messaging System using that technology. Commercial-off-the-shelf (COTS) security approaches (including encryption) from the private sector are the likely successor.

Security attempts by the private sector have had their share of problems, too. Netscape products have been 'hacked' by security professionals in order to keep security vendors honest. Microsoft is a constant target because of their historical security-apathy and denial-arrogance. Their attempts at security implementation have been marginal at best, even when they do work. Oracle, on the other hand, has developed an excellent reputation for developing secure databases that actually work.[22]

As a society, we've spent billions more building products, deploying systems - some secret, some not-so-secret – and what do we have to show for it? We've tried to build the virtual walls around our computer systems higher and higher. We've tried to make our systems impenetrable. We've tried to adapt Fortress Mentality and Risk Avoidance in a symmetrical world.

---

[22] *Application developers are perpetually plagues by security holes that constantly crop up and never seem to get plugged up completely.*

The commercial sector, as in any free-market economy has done well in avoiding some technical disasters, but still the fundamental premise of security is entirely too static. Despite excellent efforts and attempts on the part of some companies, (not enough though!) static Fortress Mentality does not enough solve the real-world problems as evidenced by increasing numbers of successful electronic assaults against systems.[23] Based upon those experiences, we now come to the following conclusions.

- As a species, we humans are not smart enough to build a computer security system that is impenetrable. The two mutually exclusive goals of protection and access sit at opposite ends of the spectrum. There is no such thing as 100% security. We are not smart enough to build a secure distributed system that will both keep the bad guys out, and let the good guys in.

- Based upon our collective experience, (a la CHOTS), if we were smart enough to build an impenetrable security system, it wouldn't be very useful or functional.

- If, we were smart enough to build a computer security system that met these goals, we couldn't afford it. So much for cost-effective security.

- Risk Avoidance and Fortress Mentality are the wrong approaches when symmetric secure communications is the real goal.

- Computer security is very, very difficult to do.

The currently popular Reference Monitor model of computer security has proven to be unworkable in the real world. So what are we to do? Refer back to the goals we outlined in Chapter 1, keep them in mind, and see how well Time Based Security fits them.

---

[23] *If we expand Bob Ayers DoD findings, with the Pentagon representing only 1-2% of all US computers, we find that there are between 20 Million and 20 Billion hacking attempts every year. Of course, different folks have different ideas on what constitutes a hack. To some, a hack is a single 'ping.' Nonetheless, the number of ongoing events is astounding.*

7
Jesse James Arrives
on the Scene

In order to start thinking like a Time Based Security expert, let's step back a hundred years or so to the late 19th Century. Train robbers. Back in the good ol' days of the Wild West, trains and stagecoaches got robbed. (Never mind we live in the Wild, Wild, Internet.) Banks got robbed, too. Gold and cash and other valuables were kept in safes. Rugged, heavy safes with big reliable locks.

Now, one of the great train robbers of the day was Jesse James. He and his gang would lasso a train, put a gun to the engineer's head, and force the train to stop. They would off the safe or other 'secure' container onto the side of the tracks and the train would be sent on its merry way. Hopefully, no one was injured in the process.

In live presentations I always find a willing victim to harass in the front row. So, I might ask him/her the following with a lunge: "Are the contents of that safe, the one Jesse just robbed and is sitting on the side of the tracks, is it secure?"

"Sure it is! I can't break in," is one answer.

"What do you mean by secure?"

"Somewhat secure."

"How secure is secure?"

Pressing the point, the answers get a little more thoughtful

"I could drill a hole through it...."

"Jesse James didn't have drills...."

"So they could blow it up...."

"And blow up the money inside ... "

"Or shoot the lock..."

"Brute force will work."

Most audiences come to the conclusion that, no, safes from the 1880's were not all that secure. Jesse could shoot off the lock, or blow it up or throw it off a cliff. Something terribly violent and terribly effective at getting at the riches inside. Bah humbug! Fortress Mentality and Risk Avoidance defeated again.

It seems that the best technology of the late 19th Century wasn't good enough to keep the likes of Jesse James from getting inside any safe he stole.

So, what good was the safe? I mean, really? What good was it? If Jesse can get into it, so can other rustlers, hustlers and thieves, using the same brute force approach. So, what good is the safe? What does it protect?

The answer is amazingly simple. The safe keeps out the good guys. It keeps the honest folks honest.[24]

The train conductor can't get into the safe unless he blows it up and is willing to take the risk. Ma and Pa Kettle probably couldn't and wouldn't either. In fact, most Hooterville residents are too nice and sweet to even attempt it, but that wouldn't stop the Cannonball from being robbed on its express to run to Pixley or the Shady Rest, would it?

Again, we have to ask – what does the safe do, then? What does it buy us? It buys us one thing and one thing only: Time.

Given that the train is robbed at 10:00AM, and it takes 15 minutes to blow up the safe, the owners of the money in the safe desperately hope that the sheriff and his posse will show up within 15 minutes; the amount of time that the contents of that stolen safe can still be considered somewhat secure.

So, here we are with a safe-full of money, safe from the good guys, but ripe for the picking from the bad guys – even if they only have a small window of opportunity – measured in time. Doesn't seem quite right does it? Fast forward. A century.

---

[24] *"Thank you Oliver Douglas and the friendly folks in Hootersville, Dahling."*

8
Fast Forward One Century

It's today. Pick a city. Any city. And imagine that in mid-town Any City, Planet Earth, there is a great, tall bank building. Inside the vast marble expanse of that bank are expensive chandeliers and floor to ceiling picture windows, polished brass and carved mahogany desks. Dominating the center of that bank lobby is an incredibly shiny vault.

The contents of this vault are protected by a mirror-finish stainless-steel door that is twelve feet high, eight feet wide, three feet thick and weighs forty-two tons. Massive eighteen-inch-thick deadbolts lock the door into the concrete and steel reinforced walls that surround the vault on six sides. If the door is closed and locked properly, is this vault secure? Think about it for a moment... are the contents of the vault secure?

Audiences typically sit back and ponder this question a bit more. That's one massively impressive vault I'm describing. People aren't as quick to judge the security of a modern high-tech, space-age-alloy vault. "Are the negotiable securities and diamonds and gold and cash secure from the bad guys?" I ask.

"Uh, yeah, I guess so..." Hesitancy from the front row. Whispers and thinking out loud throughout the room.

"Yup! It's secure. I can't break in." Arrogance from the left flank by the guy with the beanie copter.

"What do mean by secure?" Lady on the aisle, half way back. Good point.

"I could compromise the electronic locks. " Let's assume you can't for argument's sake.

"You could always blow it up..."

"With what... a nuke?" Laughter.

"Small bombs, bazooka... plastique..." The hardware store variety or the Ryder truck variety? More laughter.

29

"That's a lot of metal and reinforced concrete to get through…"

"Use an oxyacetylene torch, right?" Good.

Soon enough, led by a few of the more criminally-minded participants (no aspersions intended), the audience will jointly come to the conclusion that this mega-dollar vault in mid-town Manhattan may not be quite so secure as they once thought. We could blow the vault up or maybe burn through the walls with our oxyacetylene torches and we'd be home free and rich, right? Rob them blind? Why don't we all do it?

The audience doesn't buy my logic about how easy it should be to rob the vault. Borrow a Sherman tank and press forward!

I then ask them. "Is the vault that only thing that we use to protect the contents of that vault? Is that it? Titanium, steel and concrete?"

"There's doors to the street." A lot of good thin panes of glass will do against a frontal assault. Right.

"No, we have alarms," someone calls out.

"Alarms? What alarms? So the safes aren't good enough, eh? Don't they offer enough protection?" I ask them challengingly. Many of them are making the Jesse James connection on their own.

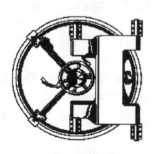 

Some bright individual will immediately say, "of course not. You have to have alarms to make it all work right." He looks around for approval and a few nodding heads give him the support he needs.

"Then what good is the safe?" I ask.

No one is quick to answer that one. Maybe a meek voice will say, "it makes the robbers' job harder.“

"Exactly!" I shout. "How much harder?"

"Er...ah...uh...that depends..."

"So what has the safe bought you?" I holler at them again.

"Protection!" Well, maybe, sorta.

This time, a few other people simultaneously answer: "Time!"

"Yes!" I dance in their general direction. "Exactly, time! Time for what?" I make them think once again.

"Time for...." Stutter.

"Ah, time to...." Hesitancy.

"Time to set off the alarms!" Right again. These guys are good.

"So, tell me about these alarms..." The audience is really getting excited now. Some are standing in the rear of the room or pacing and thinking. Others are sitting on the back of their chairs. The room is alive with brain cells firing their synapses in rapid succession.

"The alarms detect...."

"The bad guys, right?"

"What kinds of alarms...?"

"Heat sensors maybe...?"

"Sensors, yeah. Cameras? Motion sensors?"

"Microphones... trip wires...."

"Odor detectors to smell the bad guys!" someone offers enthusiastically and everyone enjoys a good laugh.

"Right, right and right again," I say parading back and forth as everyone is shouting all at once, now. We all smell satisfaction in the air. "Detection is the key here. Detection."

"So, OK, great. You got your sensors...but so what? Your bank is still getting robbed." I wait for answers. Nothing. Blank faces echoing the cranial wheels turning. "So, all we need to secure the contents of that great vault are thick titanium walls and an alarm system. Is that really all we need?" C'mon guys, we're so close!

"You gotta arrest them." A voice comes from the rear near the coffee urn. Sure, arresting bad guys is good.

31

"Yeah, catch 'em in the act."

"You call the police…"

"Or rent-a-cop."

"Goofy guys with guns!" Lots of laughter.

"And what do we call that?"

"An hour's wait…." More laughs.

"Response…."

"Reaction."

Good. Real good.

In the case of the protecting the bank's assets, the contents are first shrouded by the vault (Protection to some extent) and then an alarm system acts as a Detection System, and then the goofy guy with the gun is the Response or Reaction to the Detected assault against the Protection mechanism. It's really that simple…conceptually. Of course there is an infinity of possibilities for security configurations; the bank might use motion sensors only, but inside the vault, a heat sensor is used. An alarm might be silent or loud, and the kinds of reactions might differ depending upon the time of perceived attack. In any case, it's the same model.

Now we have all of the pieces we need to construct a workable security model that can be applied to information systems, networks and critical infrastructures. We will use these components to begin building the Time Based Security formula. (No conniption fits, here, OK? This is super-simple math my third grade son, Adam, would enjoy. No panic attacks or Valium needed.)

So, suck it in, and please turn the page!

The PDR Security Formula

9

I've spent a lot of warm-up time convincing my audience that they now have the necessary pieces they needed to create a real mathematical formula by which to view computer security, and maybe even, protect our systems and information. The three fundamental bits are:

- Protection (The bank vault)
- Detection (The sensors and alarms)
- Reaction (The cops)

Many people think at this point that I have oversimplified it. "Security is a lot harder than that!" They challenge, and I agree.

"Yes, security implementation can be terribly difficult. Agreed. But, ask yourself what model you are currently following." No offense intended at all, but very few folks have a real idea of the theory and reason behind the products and processes that they implement. Most audiences nervously shake off the question, understandably so. "By the end of today, you will have a model to follow."

The front row might nod, and the back row can hide – for the moment. "OK. What do these three items all have in common? What else do we have to work with?" I ask them.

Most people think hard...they're not yet used to this sort of security thinking. "Ah...er...um..." I hope to be pushing the envelope here. Remember, a lot of security people have been hired with those two magic words, "hey, you!" and I am trying to shake them into realizing that security is as fundamental a component in information systems, as the weak and strong forces of nature complement electromagnetism and gravity.

"We have time," proclaims a geeky guy in the front row wearing an oversized Disneyworld shirt and Birkenstocks with white socks.

"Yes! Yes! Time! Righto! And so, who sees the answer? Anyone get it? Anyone see how these all fit together yet?" A lot of intense thinking and discussion commences.

"Seems like you gotta have fast detection and reaction." This wasn't the geeky guy this time; it was a manager in the back row.

I shouted with glee! "Now you can have the formula! You figured it all out." And as I write it down on the white board, which is projected on a large screen to hundreds of attendees, I talk.

"Information security can now be viewed as a simple affair. You may not have to build up huge levels of protection any longer in every situation. You can concentrate your efforts and your budgets on something totally different: on detection and reaction... which ultimately determines the amount of effective security you have."

Pt > Dt + Rt

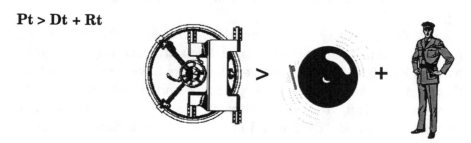

The amount of time offered by the Protection device or system, 'P-sub-t' must be greater than the amount of time it takes to detect the attack, 'D-sub-t' plus the amount of time it take to react to the detection, 'R-sub-t.' That's it.

Now stop reading for a second and think about the incredible simplicity of this argument. If the amount of protection time you provide is greater than the sum of detection and reaction time, then your systems can be considered secure. [25]

Now, think about the formula backwards. If the Detection and Reaction mechanisms you use in your networks are very fast,

---

[25] *There are caveats and other considerations that must be taken into account such as your detection and reaction systems actually work. The following chapters will look at this in much more detail.*

then you don't need as strong a Protection mechanism. Same thing, just in reverse.

Yes, it is that simple, conceptually. With Time Based Security, we no longer care how great a protection device we've bought. We don't necessarily care how high a wall we've built around our systems or how 'whiz bang' a firewall we've purchased. That becomes a secondary consideration.

What we now care about is monitoring system activity, user behavior, detecting anomalous out-of-bounds occurrences and then responding to them as quickly as possible. Some manufacturers design and sell Intrusion Detection Systems and audit analyzers which fit very neatly into the TBS model; more about them later.

Let's try an example. Assume that you have an intrusion detection system that is capable of detecting certain types of attacks, for argument's sake, in 1 second. Then, let's assume that you have a reaction system that can divert or otherwise thwart the attack to your satisfaction in 5 seconds after it has been detected. In this case:

$$Dt = 1 \text{ second}$$
$$Rt = 5 \text{ seconds}$$

For your information resources and assets to be considered secure, then **Pt** has to offer a minimum of 6 seconds of protection.[26] So your job, in this instance, would be to implement a protection mechanism (process, product or both) which offers at least 6 seconds of active protective defense. (I know, I know – how do you do that? Keep reading. We'll get there step by step.)

The detection and reaction mechanism are serial in nature, not parallel. There can be no reaction unless something has been detected first. These should be considered as 'sister processes' which function independently, with distinct rules and behavior, but which must operate together to be effective.

Another way of looking at it is:

---

[26] *For the rest of this book, I will no longer use the 'sub-t', the subscript for Time, since all terms, P, D, T and other variables will be measured in Time.*

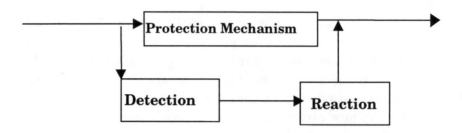

Architecturally, the detection mechanism must be placed parallel to the protection mechanism so it can 'know' what's going on in the main information/control path. The reaction mechanism must follow the detection mechanism, and be connected in some way to the protection mechanism for TBS to work. The specific type of reaction at this point is immaterial. The reactions you choose are up to you, based upon need, time-based analysis (coming up soon) and organizational policy.

Note also, the corollary to the TBS formula. If it takes longer to detect and to respond to an intrusion than the amount of protection time afforded by the security measures, that is if:

P < D + R

then, effective security is impossible to achieve in this system.

It should be becoming a little bit obvious that the choice of a good protection system is not the first thing you need to think about when designing a secure network environment. It's the efficacy of the detection and reaction processes that really matters.

Before we proceed too much further though, I want to explore the details and parallels in the physical world to help you understand the most basic premise and formula of Time Based Security.

10
Physical TBS Comparison

Let's go back to our physical bank vault to examine it in a little more detail and see how we can learn more about computer security in the process. And for you budget-impaired professionals, I have found that the following explanations go a long, long way in explaining security processes to the lay person, non-technical manager, lawyer or beancounter who may not be a fan of spending bottom-line dollars in this area.

The big stainless-steel vault represents **P**, the amount of time it takes to break through the doors or concrete walls, independent of any alarms or cops. How long is that? How much time does that represent? Here is an interesting little item that helps us in the physical world. When a bank goes out to buy a vault, or you and I go out to buy home safes, the better manufacturers will provide some specifications. Such as:

> *"This safe offers protection against an oxyacetylene torch burning at 3200 degree Celsius at 2 inches for a maximum of 18 hours."*

> *"This concrete bunker will hold secure against explosive forces up to those created by 5kg of 3.8 rated TNT."*

After that, all bets are off. At least we have a UL-style specification from which to consider our next defensive moves.

What the manufacturer has told us is how much time and or energy is required to defeat the protection mechanism; in this case the safe. At this point, we have no such exacting techniques in the world of information security, but with the adaptation of Time Based Security, all of that will change.

Next, the effectiveness of the detection mechanism, (alarms or sensors) are represented by **D**. So, the first question we need to ask is, "how long will it take the alarms to detect the bad guys?" Arbitrarily for this paper, let's say that there are several alarm systems protecting the physical bank. Each one detects certain out-of-bounds behaviors with varying times, **D**, and each of which is triggered by a specific event.

Alarms	'D' Time	Trigger Event
Door Alarms	15secs	Incorrect code entry
Window Alarms	.1 secs	Electrical Discontinuity
Sound Alarms	.235 secs	>58 db/SPL
Motion Sensors	.7 secs	.2fps
Heat Sensors	1.2 secs	95F/10secs.
Vibration Sensors	.45 secs	.32kj

So, depending upon how long the intruders take to do their job, which probably exceeds the detection times, **D**, of the sensors, their activity will be sensed. Now the last step is to build in a response, **R**. Some possibilities for response in the physical world are:

Reaction	'R' Time	Reaction Method
Loud Alarm	125 secs	Sound
Alert Alarm Co.	0350 secs	Hard wired;
Alert Alarm Co.	22 secs	Dial-up on Telephone
False Alarm Analysis	15 secs	Off Site at Alarm Co.
Call Police	18 secs	Dial-Up
Police Respond	<120 secs	Dispatcher Voice

From the available options, our hypothetical bank administrators can choose the appropriate detection and response mechanisms to implement, keeping in mind that the sum of their times must be less than the time value offered by **P**, which has been specified by the vault or safe manufacturer.

Now, let's consider a smaller facility; perhaps an office or even your home. Alarm companies will come wire up your house to 'protect' against burglars. But in reality, such companies aren't offering any protection at all. They are providing detection and reaction services. They wire your windows and doors to detect an intruder and alert the police.

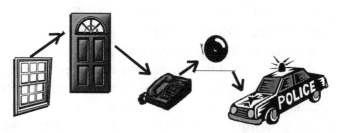

In the physical world when you want to enter your 'secure' and 'protected' office in the morning the same logic applies. You put your key in the lock and then open the door. The detection mechanism wrapped around the door (or perhaps a sonic or motion detector) now knows the door is open, but it doesn't know if you're a good guy or a bad guy. So it waits. When the system was installed, you were probably told that you have 15 or 20 or 30 seconds to run like heck to the alarm box hidden inside the coat closet and enter the correct password or code sequence. If you remember that there's an alarm, and if you succeed in entering the code within the prescribed amount of time, the detection mechanism now knows who you are (theoretically) and will turn off the reaction mechanism. (Won't call the cops.)

If, however, you forger to enter the code, or if it's a bad guy dumb enough to break into a facility with a big *Alarm and Protection Company* decal stuck on the windows, the detection device will wait the programmed amount of time and tell the reaction mechanism to do its thing. Likely, and it's happened to me countless times, the phone might ring, and an *Alarm and Protection Company* employee will call. He will ask for a verbal code, and if you're a legitimate employee, you will scrounge through your wallet or purse and nervously recite it. No cops or guards wildly waving guns will show up. Depending upon the chosen reaction mechanism, though, maybe no call will come to your office, and instead you will find yourself face to face with a Magnum 357. Explanations have to be fast... and good.

Think about protection, **P**, now. Storefronts are protected by panes of glass; negligible protection at best. (Who can't throw a brick through a window?) So, for security, stores rely upon detection and reaction systems, not protection. Thick, heavy, wooden doors (or metal shutters) offer nominally more protection, but crashing through one might trigger a sonic alarm, triggering a similar detection-reaction sequence. Residential windows and locks offer minimal protection, but the alarms and cops responding are meant to offer peace of mind. Some home alarm systems offer different detection-reaction mechanisms and options for homeowners, depending on whether they are out for the evening, on vacation or inside, tucked into bed. The rules are up to the homeowner and the *Alarm and Protection Company*.

As you are probably thinking by now, "what happens if I have no protection? What do I do?"

This is part of the beauty of Time Based Security. In the next chaplets, I will show you how to take this factor into account – and even increase your security budget in the process.

11
What Protection?

Wait. *Pause!*

Before you look at more of the (simple, simple) formulas in the coming chaplets, calm down. We're not talking calculus here! The math is exceedingly basic, but the formulas *do* represent powerful concepts and tools to effect major paradigm shifts in information and infrastructure assurance. Invest a few short minutes on the first few formulas, symbols and their meanings, and you'll soon read and use them fluently! (Now, back to our regularly scheduled program.) ☺

Let's say, as in the physical case of the glass paned store front, that we have no protection. Now we can examine the fundamental Time Based Security formula in a slightly modified way:

$$P > D + R \text{ iff } Pt > 0$$

This says that our basic premise holds true IF and ONLY IF (iff) the protection value, **P**, is greater than '0'. So if there is no protection at all, and **P = 0,** then we need to look at the situation a bit differently as we will in the next chaplet. However, if

$$P \geq D + R$$

what happens when **P** is either equal to or greater than **D + R**? Let's examine the new formula, **P = D + R.** This says that the time components of protection and the sum of detection and reaction are equal. Does this provide a secure environment? Not really.

The problem here is that while the formula may hold true in certain cases where everything works perfectly all of the time, we all know that Murphy will raise his ugly head and make a mess of things on our behalf. (Remember, even in Time Based Security, there is no such thing as 100% efficacy.)

In this situation, (**P = D + R**), the security mechanisms are given no room for error. According to the central limit theorem, if protection time is exactly the same as the sum of the detection time plus the response time, on average, security measures will only be effective 50% of the time. So although the $\geq$ operator is true, for the remainder of this paper, and in general TBS use, I suggest using only '>' in the equation. Note, also, that if it takes longer to detect and to respond to an intrusion than the delay afforded by the protective security measures, that is if the time value of protection is less than that of detection plus reaction:

$$P < D + R$$

then effective security is impossible to achieve in this system. Using the basic TBS formula backwards is worth mentioning again: If you have a detection and reaction mechanism that takes a long time to function, odds are high that your protection mechanism does not approach infinity in effective time defense; therefore, you do not have a secure environment. If we just spend a few seconds thinking about it, we can easily come to the following self-obvious conclusion.

We want the sum of $(D + R)$ *to be as small as possible; that is, we want it to approach zero-time.*

In the simplified nomenclature used here, we want $D + R \Rightarrow 0$. The ramification of that should now become even clearer. As the speed of your network's detection and reaction mechanisms increase, your need for strong protection mechanisms is diminished. Therefore, as $D + R \Rightarrow 0$, we can try to have $P \Rightarrow 0 + n$, where 'n' is some nominal incremental amount of time as suggested in the $P = D + R$ argument above. This is necessary to insure that the original formula holds, and that the amount of protection in time is greater than the amount of time it takes to detect and react to a specified event.

A couple of thoughts may have already occurred to you. What happens if there is no D and no R? Won't P be greater than their sum then? No, is the simple answer. D and R can only approach 0 ($\Rightarrow 0$) if there are D and R mechanisms in place. The desire is for them to be as short as possible, of course, but they have to be implemented to work at all.

If there is no detection mechanism, it has no measurable time component, therefore $D = \infty$ and if there is no reaction mechanism, $R = \infty$ where $\infty = $ **infinity,** or an infinite amount of time. If there is no detection/reaction mechanisms in place, then $P > \infty$ or, the amount of time that P offers would have to be infinitely large. That is the current state of the art in information security in most cases – fortress mentality. There are no detection or reaction mechanisms, therefore P must be absurdly high (where $P \Rightarrow \infty$) to have any effectiveness. And that's the whole point of Time Based Security in the first place! Make information assurance a reasonable exercise, which is measurable, and now we can see the substantiation why conventional security endeavors have limited potential for success.

	Physical Bank	Bank Time	Network Time	Virtual Networks
Protection	Vault/Locks	P' is Known	P' is Unknown	Electronic Walls
Detection	Alarms	Fast	Faster	Sniffers, Host Audit
Reaction	Cops	Fast	Faster	Email, Auto Shut Down

Comparing Physical and Virtual Defenses

Assumptions – we all know the problem with them, and in network security we face the same conundra. When we install protective devices like firewalls or encryption or any other security mechanisms, we tend to (erroneously) assume that they are working correctly. Experience tells us that this false assumption is at the basis of many of our current network and infrastructural woes. Using our worst-case approach, assuming that $P = 0$ is a much safer opening gambit, if security is the real goal; there are just so many things that can go wrong in a network.

Manufacturers will not (and at this point, cannot) guarantee the defensive performance of their products. It's a hunch and hope. TBS will hopefully change this in the coming years.

- We cannot measure the efficacy of virtual protective systems – yet.
- Networks grow every day. Their inherent dynamics change the security posture of a network on a constant basis.
- Administrators have a difficult time knowing every single network ingress and egress. Modems, PC Anywhere, unknown phone lines and secret subnets plague organizations.
- Connecting enterprise networks to partner organizations with unknown security can weaken a network's defensive strength.
- Seemingly harmless applications often innocently create security vulnerabilities.
- New hacks appear daily against leading applications, operating systems and security mechanisms. Organizations have a terribly difficult time keeping up with every new one.
- It takes time and effort to install new patches to enhance security, and they don't always work.
- Well-designed security mechanisms are all too often installed incorrectly and/or completely misconfigured.
- Administrators often turn off security controls during audits and maintenance and forget to turn them back on.
- You can't adequately test the protective value of a network with any degree of assurance beyond the exact moment it was tested.

That's enough to get the point across: conventional, protective information security is very difficult. And so, we assume for many TBS applications that **P = 0**. In the next chaplet, we will examine what happens in this case, which may prove to be one of the most valuable tools you can use today for assessing and increasing the security of your information and intellectual assets.

12
Measuring Your Security

How do you and your organization stand security-wise? How do you stack up in your defensive efforts?

Let's look at our mythical bank again. Say it is the middle of the day. The bank's big metal vault door is wide open; there is no protection at all, ($P = 0$) and the detection schema are more human in nature. (*"Hey, Joe! See that guy with the gun leaping behind the teller's cages and running for the vault. Think we gotta problem?"*)

That is one kind of detection as defense (albeit slow and human), and then an appropriate reaction is called for. The bad guy has only so much time to grab the money and escape before a reaction mechanism has been put into place to stop him. The guards might wake up and shut him into the vault. Or they might shoot. Or they might call the cops. Figure the amount of time it takes to thwart a bank robbery and you can get a feel for how TBS works.

Same sorta thing happens if the bad guy is trying to hold up the bank. "Stick 'em up!" he shouts at the teller who has an unprotected drawer full of money. "Fill the bag with cash – fast." Robbers instinctively know that all security is time-based. "Hurry up!" they holler. The astute bank, though, might have installed a Panic Button near the tellers' feet, which they can discreetly press while filling the bag with cash. Here, protection is near nil ($P = 0$) yet a reasonably fast detection mechanism has been designed, and then it's up to the reaction crews to minimize the damage and/or thwart the robbers.

Back in our networked world for a minute, ask yourself these questions: Have you ever sat down and tried to find out just how tight or loose the security is in your networks? Have you ever looked at the *Process* of defense? Now you have the chance to start putting Time Based Security to work for you – right now.[27]

---

[27] *Feel free to use the charts provided in this book and copy as many of them as you'd like for additional use. Hand them out, pass them around, fax them to friends and bosses. Just use*

In one TBS presentation I gave, the results of the quick live assessment absolutely astonished me. The presentation was for a large government organization with huge facilities all over the country. After explaining the TBS basics, I asked the following question. "What sort of detection mechanisms do you have installed in your networks, and how well do they work?" Keep in mind that the security chief of this agency was present, and is a big fan of the Time Based Security approach to infosec. To my absolute amazement, the LAN security administrator, several layers of management down, spoke up and said with pride, "Oh, we have audit trails." He crossed his arms on his chest in smug defiance.

"Do you use real time audit trail analysis?" I asked.

"No," he answered.

"And you *do* use automated analysis tools to search for anomalous behavior?"

"No, we don't," he affirmed. I am trying to help this guy and he's defending his methods.

"OK, these audit trails. How often do you check them?" The security chief was shaking her head in disbelief.

"Oh, every couple of weeks if we have time." Big ouch! An abysmal failure of security.

But let's take another example and apply the approach to your networks. For the moment, let's forget about protection, or 'P'. OK? We're going to assume that **P = 0** and instead only look at detection and reaction mechanisms **D** and **R**.. First, I must find a suitable person in the audience to pick on... er, ah... use as an example.

"How many people in the audience are system administrators?" People nervously look around because I told them at the beginning of the session that I would be picking on people, and this seems like it just might be that time. A small handful of arms are hesitantly raised. I find my victim, and a collective sigh of relief fills the auditorium.

---

them and see how much value they provide for you, and for making management understand just how difficult information security really is.

"Dick, you work at a large insurance company, right?" He won't tell us which one, and I don't pursue it 'cause it doesn't matter. He's meant to be an example, not a whipping boy.

"And you have a detection system of some sort, right?" He nods in agreement. "What kind?" He says that he uses a specific intrusion detection product, which generates audit trails in his NT, Unix and mainframe environment. This is good news and will make an excellent example for the rest of the 500 or so people in the audience.

"I also assume, Dick, that you have chosen events to detect and monitor that are unacceptable in your networks, right?" Like a good security administrator he readily agrees. The events could be three failed logons, or too many system requests of the same processes, or too many printouts from the server. It doesn't matter what out-of-bound events he has chosen. That choice should be based upon his company's policy.

I continue with Dick. "OK. Let's say some bad guys or just pain-in-the-tuckous, ankle-biting hackers are knocking loudly at your electronic doors. Maybe they're running unapproved port scans or password checkers or repeatedly pinging you... whatever. You can detect those things, right?" He humbles a bit and wiggles his flat hand with a so-so motion, meaning he can do some detection, but not all.

"Fair enough. In this case though, we'll say you can detect the annoyance or bad guys with whatever products you're using." He smiles with the improved state of affairs I have just given him.

"So, everyone else in the audience, please silently ask yourself the following question, while Dick here answers aloud. You have detection and auditing software in place. I assume you have properly configured it, and I further give your top management credit for handing you a corporate policy to enforce." Dick cringes a bit. This is an ideal scenario I am painting him, and it's clear his firm is not that security-aware yet. "Whether or not that's really true, let's give Dick here credit for having all of these things in place. Let's be as fair as we can here, OK?" Everyone agrees that's fair.

"Given these generous assumptions," I say smiling, "from the time that an ankle-biter begins nibbling, how long does it take your detection mechanisms to detect the inappropriate behavior?" Dick's eyes roll up toward the ceiling in mental calculation.

"Ah, I should have a notice on my desktop, in, oh...say ten to twenty seconds."

"You get an email or an alert at your desk?"

"Yup."

"In about 15 seconds?"

"Yeah, I guess. We've never really measured it...." And Dick's mind is beginning to shift to a Time Based Security viewpoint.

"So let's be generous again and say your detection process takes 10 seconds, OK?" I offer the reduced time and Dick readily agrees. "So in Dick's case, **D = 10** seconds. Now, let's look at Dick's '**R**'. Dick, how does the detection system notify you that something is wrong?"

"If I'm at my desk, then I know pretty much right away..."

"And if you're at lunch?"

Dick slithers down into his chair slightly. "Ah, it's takes longer. I'll be paged."

"And how long does that take?" I push the point... again, only caring about time as the common security element.

"Ah... say five minutes."

"And if you're at home? Or at dinner? Or it's the weekend?"

He shakes his head in good humor. "Hours... days!" he says now understanding a bit more of how the reaction component affects Time Based Security.

"There is a solution to that at-home problem, though," I suggest to the audience and Dick. "A friend of my at the Department of Justice thinks it's a good idea. On all web sites and log-on screens, put up the following message:

> *Welcome to Our Company. We are sorry, but Dick, our security administrator is home with his family for the weekend. All hackers and on-line miscreants are respectfully requested to restrict all hacking attempts and attacks to between the hours of 9-5, Monday thru Friday.*

"So we have a detection time of 10 seconds, another 2-3 minutes to let Dick know something is wrong and another 5 minutes to 5 hours for him to get to the point where he can respond to the situation." Dick nods, cowering in mock defeat, and the audience groans with sympathy pains.

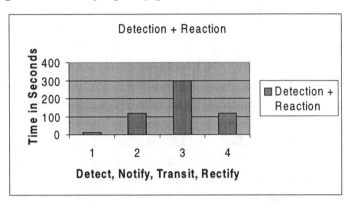

"And," I continue, "we still haven't fixed the problem, have we? Once you're at your terminal, Dick, how long does it take you to identify, corroborate and rectify whatever situation or anomaly has been detected?"

"Oh, that's pretty quick," Dick says brightly. "Usually just a couple of minutes."

"A couple?" I challenge.

"Ah, yeah, maybe five minutes..."

"OK, we'll give Dick the benefit of the doubt at say it only takes him 2 minutes." I glare at him and he grins back. He's a real good sport.

"Therefore, in the best case, we have 10 seconds, 2 minutes, 5 minutes and 2 minutes." Dick nods. "That's 9 minutes and 10 seconds." Everyone agrees.

"That's pretty impressive," I tell Dick. "One Department of Defense audience member once told me they have nothing of value therefore they don't bother even protecting it." Laughs.

"So, now let's ask you this. Assume you have something of value on your networks or your company cannot function without them. Now assume you have a naked network. A network where the defensive perimeter may have been compromised, or it collapsed, maybe there is no perimeter defense at all. Given a reasonable adversary who has done his homework, and knows his way through your networks (or maybe it's an insider gone bad) in 9 minutes and 10 seconds of unrestricted access, how much damage can this adversary do?"

Tough question, and a tougher one to answer honestly in public. Dick thinks for a few seconds, all ears at the ready, and he finally says, "we're screwed."

I offer the highly technical term for this condition: S.O.L. which relieves the tension a bit, but the point is made. "What do you think your adversary can do in a denial of service attack in 9 minutes or so?"

"We'd be essentially out of business," Dick admits. "We have to have the internal networks up, and we use the Net more and more for communications with our agents and representatives... a denial of service attack pulls us off the playing field." He was right. Absolutely right.

"And what about the information-rich treasure trove; the bits and bytes goodies. What are those?"

"Our marketing plans, pricing, profits, customer lists, medical conditions... we have a lot to protect."

"Again, in nine-plus minutes, what could your adversary do?"

"Steal it all."

"All of it? Really?"

Dick thinks for a minute, and asks, "time, huh?"

I nod.

"We have a 10 meg network, and a T-1 to the outside world, so how much data can be transferred over a T-1 in nine minutes... is that the answer?"

Exactamundo! Perfect. Dick could not have been more right in his answer.

Dick first needs to know the bandwidth usage history of his T-1 and his internal comm lines to determine both average and worst-case situations. So, the T-1 bandwidth is 1.54 Mbits/sec X 550 seconds (9 mins/10sec) and the worst case (assuming Dick is working at maximum speed, is nearby his desk and he can fix things quickly) is 847 Mbits of data, or approximately 100MB. (In the next chaplet, we really begin to drill this concept.)

"What about if the goal of the attacker is not to steal your 50MB files, but to put you out of business... at least for a while. How much damage would a sustained denial of service attack cause?" Again, Dick cringes.

"A whole lot... nine minutes is a long time." He's right. Denial of service scripts tend to operate very, very quickly, and can be launched over and over again within a very short period of time. (See the chaplet entitled, "Surviving Denial of Service.")

"Fine, we all agree that DoS attacks are both very quick and very bad. Let's say that you are under a sustained attack... of any kind. Theft, integrity or denial of service. Dick, let me ask you this. Are you permitted to shut down your entire network when you feel it is necessary without having to ask someone else permission?"

"No!" He retorts quickly. "I have to talk to the VP of MIS to do that...."

"No problem," I add. "How much time does it take then for you to contact the VP of MIS, explain the situation and get approval to shut it all down."

"Uh... if I'm lucky he's not in a meeting ...so I guess five or ten minutes..."

"And if he's at lunch....?"

"Longer...."

"And if he's on travel...."

"Even longer."

So, Dick has just discovered, that the generous nine minutes and ten seconds he had as a (**D + R**) has just grown by 50-100% if he has to take some drastic measures to thwart or remedy the situation. Ouch.

By working through the above discussion and answering the following questions, you and your management can assess one aspect of your current security posture; that of vulnerability as a function of time. Try working these steps for a small portion of your corporate networks first, to get a feel for how the process works, and then expand it to your entire enterprise. Of course, as you learn more about the Time Based Security process, you will be able to add other pieces based upon function and criticality rather than considering all components equal. The following questions assume, as before, that $P = 0$.

1. Do you have any detection systems in place? <u>Yes/No</u> If you have no detection component, then $D = \infty$, and assuming $P = 0$, then $E = \infty$ and you have a virtual 100% system exposure. (See the next chapter for how E fits into TBS.)

2. Do you have a Reaction mechanism in your network, which is triggered by the detection component? <u>Yes/No</u> If you have no reaction component, then $R = \infty$ and assuming $P = 0$, then $E = \infty$ and you have a virtual 100% system exposure. (See #1 above.)

3. When your detection mechanism is optimally running, how long does it take to detect a particular attack it is designed to attack? _____ (Sec/Min/Hour) This is a tough question. As of today, vendors are not in the habit of specifying performance in time units. In a later chaplet, we will address the role of security vendors and Time Based Security. You may need to estimate this answer until measurement tools and benchmarks are developed... which won't be too long.

4. Once your detection mechanism has detected an attack or other out-of-bounds behavior, by what means are you notified? (ie, email, pager, phone, not at all, future audit trails?) _____

5. How long does it take to notify you in Step #4 above?
   - When you are sitting at your desk? _____ (Sec/Min/Hour)
   - When you are at lunch in the building? _____ (Sec/Min/Hour)
   - When you are taking a break? _____ (Sec/Min/Hour)
   - When you are headed home? _____ (Sec/Min/Hour)
   - When you are at the movies? _____ (Sec/Min/Hour)
   - When you are sleeping? _____ (Sec/Min/Hour)
   - When you are at the kids' softball game? _____ (Sec/Min/Hour)

6. Once you have been notified of a problem via the detection mechanism, on average, how long does it take you to get to a location where you can do something about it? (Or get someone else to begin corrective measures?)
    *   When you are sitting at your desk? _____ (Sec/Min/Hour)
    *   When you are at lunch in the building? _____ (Sec/Min/Hour)
    *   When you are taking a break? _____ (Sec/Min/Hour)
    *   When you are headed home? _____ (Sec/Min/Hour)
    *   When you are at the movies? _____ (Sec/Min/Hour)
    *   When you are sleeping? _____ (Sec/Min/Hour)
    *   When you are at the kid's softball game? _____ (Sec/Min/Hour)

7. Once you begin to solve the problem, on average, how long does it take you to determine the cause and effect an acceptable remedy? This might include coordination with other admin folks.
    *   When you are sitting at your desk? _____ (Sec/Min/Hour)
    *   When you are working from a remote site? _____ (Sec/Min/Hour)

8. If the attack is severe, and you are not empowered to take any and all steps necessary to protect the network, how long does it take you to get permission to take the worst case action: shut the entire network down in order to protect it? _____ (Sec/Min/Hour)

Now, please add up the *best* case numbers from the above answers: _____ (Sec/Min/Hour)

Now, please add up the *worst* case numbers from the above answers: _____ (Sec/Min/Hour)

E = _____ to _____
     (Best Case)   (Worst Case)

This is your range of **E**, or *exposure time* in your enterprise based upon a Time Based Security Analysis.

If you have multiple paths throughout your networks, take a look at each of them, fill out a separate chart for each, and compare answers. In larger networks, different administrators have different policies to follow, they are empowered differently, and the systems they are charged to defend have differing levels of mission criticality. So, piece together aspects of your networks as standalone sub-networks before trying to glue them all together into a single whole picture. Some networks are so complex, topological simplification becomes a strategic security goal; not more technology – less!.

In our example above, you can estimate how much time an adversary could possibly have to run rampant through your networks, and the answers are in all likelihood not too terribly pleasant – especially if the time values creep up to an hour and beyond. This simple quantitative analysis can help you convince management that you have a problem, and you need additional budget dollars in search of solutions.

But in an effort to further quantify security, we need to examine just how to measure the potential for damage in a way that you may never have previously considered. Please sit back and take a break before moving on to the next chaplet.

13
Measuring the Effects of Exposure Time

Since Time Based Security is also about quantification of security and risk analysis under a single umbrella, let's see how much information we can derive from what most people already know about their networks. As you are learning, it's not so simple any longer to merely say, "we're vulnerable." TBS provides the framework to force a firm to really define what it means by vulnerable; how vulnerable is vulnerable? What exactly is vulnerable? Which security fundamental is being exploited? What is the real risk?

In the previous chaplet, Dick the sys-admin guy said his company ran a 10 meg internal network, and a T-1 to the outside world. We asked the question, how much data can be transferred over a clean T-1 connection, and the answer is roughly, 1.54Mbit/sec, which once we are all through with overhead, error correction, packet buffering and the like, translates to an approximate average of 154KB/sec.

Since we are concerned with both best and worst case analysis to determine the maximum damage that can be caused by an intruder, let's ignore the bandwidth hogging variables, which seemingly work in our favor. Instead, let's work with a rule of thumb: bandwidth measured in bits divided by 10 is the maximum bandwidth in bytes.

BW/bits / 10 = BW/bytes

So, a T-1 is then about 154KB per second, or 9.2MB per minute. Now we have a method to add another quantitative element in our security assessment. How much time is required to steal information? Despite the hype of instantaneous-light-speed communications, nothing is really instantaneous. So let's develop another TBS idea, and maybe a usable metric:

File Size/Bandwidth = Required Attack Time
> or

MB/Mb/Sec = (Attack) Time
> or

F/BW = T

If the attacker's goal is theft of information, the size of the critical target files, (**F**) and the amount of critical information they contain, divided by the maximum bandwidth of the communications path (**BW**) determines the amount of unhampered attack time required, (**T**) and thus is one measure of risk. An information theft can now be measured using time as the quantifier in addition to intrinsic value.[28]

This assumes, though, another worst case: that the attacker knows the precise routing to the desired information repository, and can get there in '0-time'. This is not as unrealistic as it may sound at first, since so many modern attacks are pre-scripted. While 0-time is not real (instantaneity is still science fiction), we are often talking about attacks only a few hundred milliseconds to several seconds in length.

We can also use this same formula to find which information-rich files are of a particular vulnerability. If the maximum bandwidth of a comm path is 10Mb/sec on the internal networks, yet all external communications are channeled through a single point of a lesser bandwidth, such as a T-1 connection, we must consider the "weakest link" or "bottleneck bandwidth" as the metric by which to measure our exposure for an external attack. Internal attacks can send information out of the networks with the same TBS restrictions, or they can use a floppy or tape. In this case, it's the 154KB/sec T-1 circuit.

Back to the basic Time Based Security formula. If a particular network has an Exposure Time **E**, (**E = D + R**) and a bottleneck of T-1 speed, equals 10 seconds, only a single file in

---

[28] *When hackers broke into Pakistan's nuclear program computers, they said they stole 5MB of data in 14 minutes, for an average transfer speed of 6KB/sec, or about ½ an ISDN channel. Of course, we don't know how much time they spent poking around versus downloading, but you get the point.*

excess of 1.54MB or many files whose total size equals 1.54MB could be stolen in their entirety, before the detection/reaction mechanism takes hold and scuttles the attack. In a bottleneck of high T-3 or ATM speeds the actual potential for loss is much greater, and similarly, if ISDN speed is the slowest comm channel, the risk is substantially less, because the bandwidth is about 1/10 of that of a T-1.

Under TBS, we begin to understand that by their very nature, high-bandwidth channels increase the vulnerability to a company's networks and electronic assets. Thus bandwidth limiting in certain network paths is a valid security method. Again, non-security oriented security that is highly effective. If a particular LAN requires minimal access, then bandwidth squeezing at its access point will limit the effects of confidentiality attacks. Several approaches to adding security in a working TBS architecture become apparent, although they all won't probably be feasible until the technology is developed and marketed.

Data Padding Buffering. Assume that a company has a working TBS defense, and it has carefully constructed its $E = D + R$ (Exposure Time) to be no more than 30 seconds.[29] In this case, the minimum bandwidth (bottleneck issue) is T-1 speeds, and they are clean lines. Therefore, the maximum risk is

File Size/T-1 = 30 Seconds

File Size = 154KB/sec X 30 Seconds

F = 4.6MB

In this case, all of those critical files to be protected should be padded to a minimum of 4.6MB in size. Thus, any attacker will be able to download/steal only a portion of the target files.

The above approach has weaknesses, though. Perhaps even only portions of the target files are useful to the attacker, so in this case, file buffering only further impedes the attacker, but does not fully protect the contents. So, it makes sense to pad or buffer the front end of the files, leaving the 'good stuff' to sit safely at the back-end of the files. Time Based Security says that in this way the attacker has to wait for the entire file to be accessed/downloaded

---

[29] *This does not mean that it has not attempted to build in protection products and procedures; we just aren't willing to assume that they work or are configured correctly.*

before anything of value is lost. Other file buffering/padding may work too, in those cases where all of the data in a particular file is required, and any piecemeal theft is worthless to the attacker.

- **Integrity Wrappers**. Wrap the files of interest in a cryptographic shield, which can only be decrypted with the correct keys. Of course, this relies upon better key-management and distribution schemes, several of which are now appearing on the market. Even if the integrity keys are stolen, too, and only a portion of the target file is stolen, the attacker will no longer be assured of the accuracy and validity of the data.
- **Encryption**. Like the Integrity Wrapping approach, the entire message is encrypted, and the entire message must be available for accurate decryption.

Encryption and Integrity wrapping are powerful protective technologies, which have both pros and cons, and we will examine in greater detail later. However, they do provide relatively simple means for enhancing the defensive posture of valuable electronic assets. Bandwidth Measurement for network and information vulnerability assessment provides another metric to quantitatively demonstrate the need for additional security measures.

Integrity attacks provide different answers in Time Based Security. The attacker's goal is to make unauthorized changes to files or records – undetected, of course. The nature of such attacks does not require the complete download of the target information resources, so the means to measure vulnerability is slightly different. We want to determine how much effort, in time units, it will take to make an integrity-mess of your files. For this we must:

1. Assume you are a company insider with physical access to the networks, and that you are logged on to an inside computer (of whatever type). Measured in time, how long does it take you to manually navigate to the target application. _____ (Sec/Min/Hour)
2. Assume the same as # 1 above, but execute a script written to automate the task. How long does that take to occur? _____ (Sec/Min/Hour)

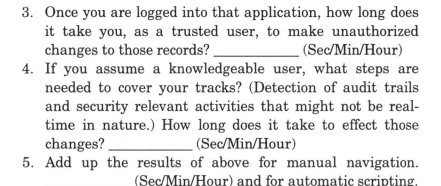

3. Once you are logged into that application, how long does it take you, as a trusted user, to make unauthorized changes to those records? _____ (Sec/Min/Hour)
4. If you assume a knowledgeable user, what steps are needed to cover your tracks? (Detection of audit trails and security relevant activities that might not be real-time in nature.) How long does it take to effect those changes? _____ (Sec/Min/Hour)
5. Add up the results of above for manual navigation. _____ (Sec/Min/Hour) and for automatic scripting.

These answers provide a range of reasonable guestimates as to how much time is required to launch an insider's integrity attack. You can perform the above steps for critical applications and data throughout your network. In each case assume that you are a knowledgeable insider with access to those target information assets. This is a somewhat empirical approach, and will be different for each organization that wishes to evaluate its own Time Based Security weaknesses and benchmarks.

Once you have the insider's view and have quantified the internal risk as a function of time, those figures can be used as a target *maximum* for the value of **E**, and thus provide a guideline and target for the desired detection and reaction speeds.

Lastly, let's take the view of a dedicated outsider, and in the worst case analysis, assume that he has gathered enough information about your networks to be considered knowledgeable. The bad-guy/gal could have gathered this information with the cooperation of an insider, through technical surveillance, social engineering, dumpster diving or any other method available.

You, as the sys-admin for security would like to know how much time it takes to launch an external integrity attack. (Design whatever attack you use to accomplish whatever nastiness suits your purpose.) Your job is to script an automated attack along with an associated time clock to measure the effectiveness of your efforts. Network based attacks are increasingly being scripted to increase the number of attacks that can be launched in a short period of time. Remember the automated attacks against the

Pentagon in early 1998? Hundreds of thousands of automated quasi-attacks in a couple of days. The bad guys also want to speed up the attack process to minimize the likelihood of getting caught. Don't forget, the bad-guys have instinctively understood Time Based Security since the dawn of bad-guys. The rule of thumb now seems so obvious: *"Don't hang around waiting for the cops to get you."* Now it's our jobs to lessen their advantage and do the same thing.[30] The first two attack approaches described above parallel two of the fundamentals of information security defense:

Theft = Confidentiality breach, and

Data Diddling = Integrity Attack.

Now it's time to look at Availability, or Denial of Service attacks. With DoS, though, time seems to be terribly sped up, and as humans we can't so easily fathom time measured in microseconds. Thus, applying Time Based Security to DoS is the toughest of all. Scripted DoS attacks occur at very high rates of speed where a complete assault could take place in a few milliseconds. While detection and reaction are critically important in solving Denial of Service attacks, we must add a new element not found in most networks to work through the problem. In the chaplet, *Surviving Denial of Service*, I will introduce derivative Time Based Security methods to defend against these cowardly, harassing attacks.

In the meantime, let's take what we have learned about the quantification of network weaknesses in the time domain, and see how to further convince management that a) you need additional budget and staff dollars to adequately defend against the bad guys, and b) Time Based Security approaches to defense are a significant valuable contribution to your overall defensive efforts.

---

[30] *If TBS is the defensive posture, then 1/TBS is the offensive posture. InfoWar Attack scenarios for the DoD and intelligence community follow the same logic. Attack scenarios need to be designed with time as the critical metric for success. Just as anti-terrorist groups time their activities down to split-seconds (to save hostages, perhaps) IW-Attacks are more successful when "you get in and get out fast." Conventional physical military insurgency operations function in this paradigm. We are currently expanding the TBS model and developing these techniques for offensive operations.*

14
Exposing $

Earlier we blew up the door of the mythical bank's vault with a high explosive. Then, we let the bad guys grab as many stock certificates and as much cash as they could carry and run like the devil before the reaction police arrived on the scene.

Again, let's assume that the bank has no protection mechanism at all. Maybe the vault door is wide open as it was in our previous example. However, to deter crime, the bank has wisely put a number of well-designed detection and reaction components in place. The **D** and **R** are well defined, measurable and replicable. The question we now ask is, given **P = 0** and (**D + R = E**), at the very worst, "How much damage can the intruders cause in the amount of time it takes to effect (**D + R**), or **E**?"

In our previous physical bank example, when we add the **D**'s and the **R**'s together, we found the bad guys have got less than 3 minutes to make off with as much loot as they can before the cops arrive. The bank then has to determine quantitatively, based upon the known contents of the physical vault, how much, in terms of dollars, that worst-case risk represents.

With Dick, our sys-admin earlier, the bad guys had even more unfettered time on their hands. I asked him, if you recall, "Given a reasonable adversary who has done his homework, in 9 minutes and 10 seconds of unrestricted access, how much damage can this adversary do to your networks?"

Dick's answer was, "we're screwed," to which I amended the technical term, S.O.L. ☺ With the tools of Time Based Security, you have evaluated your network's worst case vulnerability using two different approaches.

E = D + R and **F/BW = T**

Therefore, if we know the value of **E**, we can back into the value of **F**, by assuming **E = T**, or if we know **T**, we assign a target value to **E,** (and the target sum **D + R**).

Part of the TBS exercise in the last chaplet was to identify the size of the mission critical files, which are of special importance to your company. (I.e., formulas, customer lists, pricing, strategies, schematics, competitive bids, etc.)

What we don't know yet is the real, hard dollar cost to the company. Here is where the Security-folks have to coordinate and play nicey-nice with the Auditors and Management, and ask a series of TBS-ish questions. Try asking the following hypothetical questions (or ones of your own making) of the right people in your organizations.

- "Using TBS, and based upon current network performance and security systems in place, we have quantitatively determined that the following information and resources are at risk if we cannot rely upon our protection systems. (Provide the list as best as you can now.) If one of these files (or multiple ones) gets out, into a) the wrong hands or b) the media, (for example), what will be the financial effect on the company?"

- "We have found that some of our network components are vulnerable to specific Denial of Service attacks, (list them) and we have proven the attacks' effectiveness (describe how). The effects of these attacks range from a) shutting down the non-revenue-producing web site, to b) putting our company out of communications with our offices, partners and customers. What is the hourly/daily cost to our company if this occurs?"

- "We have determined that by using such and such methods (describe them exactly), the following critical files can be altered without authorization by both insiders and outsiders. What is the legal culpability of our company if a client's records are compromised? What is our legal status if an employee's records are illicitly accessed and subsequently used against that person?"

These exercises do presume two things, which are outside the scope of the TBS model itself, but critical to any serious quantitative efforts at improving the defensive posture of an organization.

- The organization must have some idea as to *what* information resides within its networks, and hopefully some idea as to *where* that data resides logically and physically. (If you don't know, then you might just have a serious management and organizational problem.)

- The company's Auditors have to have agreed upon some qualitative method for determining the intrinsic and extrinsic value of the company's mission critical information. They should be able to estimate how the operation, revenue, profitability or image of the company is affected if that information is compromised, lost or made unavailable. Unfortunately, in a large percentage of cases, this exercise is not conducted and these numbers are not known, so best guestimates must be substituted.

Clearly, as in our discussion on Fortress Mentality earlier, we cannot and do not want to be in the Risk Avoidance business: it doesn't work in the symmetrical world of open commerce. Risk Avoidance is anathema to our goals, and besides, we can't afford it. We are, however, in the risk management or risk mitigation business and Time Based Security offers a rationale way to quantify the risk, ultimately in hard dollars and sense.

Your honest answers to the above questions will then provide you with a very clear direction for security budgeting and implementation. If the information evaluation comes in at $1Million, for example, then the corporate policy, in conjunction perhaps with insurance underwriters and legal counsel, need to decide how much financial effort is reasonably assigned to defend that asset. That decision is not the job of the information security professional. Remember, we are in the business of keeping our client's business thriving; our job is to keep their iterative business cycle from being interrupted. Be careful not be sucked down the road into areas where your expertise is nil. Manage the process, not the answers.

Let's set up a couple of hypothetical examples where there are two seconds of naked network vulnerability and see how this process can work for you.

Assumption: **P=0** and **D + R = E = 2** secs. (100% network vulnerability for 2 seconds)

Case 1. The intruder has done his homework, and understands your internal network 100Mbit structure and connectivity. He has assembled maps of your networks, its barriers and detours, and has created an automated script for navigation and penetration through your external T-1 connection to target your intellectual valuables. He knows what he wants, where it is and how to get there. Theft of that information is the goal.

In two seconds, what is the most that he can achieve? Is his task/goal achievable? What is the actual dollar loss to your organization if he is successful? Is the information of such value that perhaps it does not belong in your electronic domains? What is the level of appropriate protection to combat this scenario? Try using the formulas from the last few chaplets to find the answers.

Case 2: The adversary wants to shut down your electronic services, say web based banking, with denial of service attacks. Your detection/reaction schemes ($D + R$) allow full recovery from such an attack in less than 2,000 msecs (2 seconds). What is the cost to you if such an attack occurs once a month? Once a week? Three times a day? Once every ten minutes? Again, a cost based analysis must be used to determine if the risk is acceptable. Should additional protection be implemented, or should the detection/reaction mechanism be enhanced to further reduce the exposure?

Case 3. You determine that your current $D + R = E = 120$ seconds in a T-1 environment. Your Important Data Base files come in 76 12MB segments, each worth $3.4 Million, and in 143 1MB pieces worth $500,000 each. What is your maximum exposure in dollars? What will you have to do to keep your adversary from getting even one complete small file?

As you think through these problems, a thought might be occurring to you. How much is data really worth? That age-old question is covered in the next chaplet.

15
Data Valuation

1. *How much is a five bedroom, ranch style house with pool on two acres and a private lake worth?*

2. *How much is a 1931 Bugatti automobile worth?*

3. *How much is the world's oldest tube condenser microphone worth?*

Answer: Whatever someone is willing to pay for them.

In the real physical world of commerce, the free market economy does a pretty good job of establishing value; both perceived and real. Perceived value is how much folks think something is worth, and real value is how much hard-earned cash they shell out to actually pay for it and take it home.

However, as economist Joel Kurtzman points out, we are in desperate need of an economic model which allows us to measure the value of that great intangible: information.

For example, on how many financial filings of a business's activity have you seen line items which say, *"Information Assets."* Hard physical assets are there, they can be touched and depreciated – we have a system upon which most of us agree. Some balance sheets include a line item for *"Good Will"* which is sort of crystal ball gazing at its best. But specific information asset evaluation is very different and much harder to handle.

"Information is the only asset that can exist in two places at once", my friend Charlie Robertello said to me one day and I have never forgotten it. I consider it to be axiomatic. Even electronic money doesn't have the luxury of being in two places at once due to the security controls imposed upon it.

Assume I have a diskette or CD-ROM with information that is worth $1 Million. A buyer has agreed to pay me that amount for it, which pretty well establishes a value. Now, I make a copy of that disk. How much is each one worth? Are they both worth $1M each;

have I doubled the value? Or are they each now only worth $500K; have I halved their value but maintained a constant total value? A copy of any kind of software falls into this conundrum. A single piece of software is worth the sum total of what people are willing to pay for it, nothing more, nothing less.

But what if the $1Million disk ends up in the hands of a competitor to my buyer? What is it's value then? Is the information really worth any less? Or is it worth the same sum total to the free market economy, but distributed amongst different people? We don't have all of these answers yet, so a strictly quantitative approach to data valuation may not be possible. We need another mechanism.

The value of Information is further confused by two competing and diametrically opposed forces: entropy and anti-entropy.

Information entropy occurs when information loses value over time. A big secret may have some discreet value associated with it, but once it leaks out, it is worth less. Once it becomes common knowledge, some or perhaps all of its intrinsic value has dissipated. Such is the case with product planning, car design, financial reports, endings of Hollywood scripts... and so on. So, in this case, the value of information is highly dynamic and entropic in that it dissipates over time.

On the other hand, information is also anti-entropic. That is, the value may still be dynamic and changing, but it is increasing in value over time. Some examples of this are "ideas ahead of their time, waiting for the right time to be introduced." The source-code of a product that becomes very popular goes up in value, as does the value of a company which is successful. As the finishing touches on designs are made, the value goes up, as does a well-honed customer list or a much appreciated piece of art.

I do not claim to be an economist, but these principles do guide a great deal of our economy even though we may not look at them in this way yet. The question we face, though, when applying Time Based Security, is "how do I know which information/files/resources are really worth something, so I know where to place my budget dollars wisely." Great question, and a tough one to answer, for which I will resort to the fine art of Approximation.

Rather than allowing the network to be viewed as some vaporous whole, which magically stores and provides information on demand, organizations need to think through and assess the real value and mission criticality of its information assets. The following charts can serve as the first step in isolating the electronic keys to the kingdom from the worthless clutter.

I like to divide all information assets into 4 broad categories for simplicity's sake. (Horizontal across the top of the chart.) If you would like to use more information categories, that's fine. This chart is not an absolute, it is a tool to be modified as you see fit. Further, the nature of vulnerabilities in any organization fall into three broad categories: Confidentiality, Integrity and Availability (Denial of Service.) Any analysis of the value of data must consider how it is to be abused by the potential adversary. Thus, if you examine the data in all three manners, you will get more accurate representations of its true value to your organization.

- **Company Proprietary:** This is the information that allows the company to differentiate itself from its competition, or which provides the basis for value-added products or services. It could include designs, product planning, customer lists, pricing strategies, discoveries, patents, source code, bids... every company has such a list of information with varying degrees of value.
- **Private Employee:** The files on your employees are exceedingly private, often strictly protected by law, and if abused, the company can find itself in an unpleasant defensive posture that can quickly get very expensive. Salaries, medical histories, performance reviews, written warnings or admonishments, social security info, family members, etc. are all in need of protection.
- **Customer Private:** Your customers' history with your firm, purchasing info, pricing, plans, and especially that information that was provided under non-disclosure. This sort of information requires protection to prevent

accidental or intentional release and to maintain client confidentiality.

Date				
Location				
Server				
Confidentiality: **If this data is released**	Company Proprietary	Employee Private	Customer Private	**Business Partner, Government,**
The results will be absolutely disasterous with no chance of economic or politcal recovery.				
There will be severe financial, political or other undesirable results, but we will survive.				
It's gonna cost us big time, but spin doctoring will take care of it.				
Negligible effects, but we still really don't want it to happen.				
Publish it all you want. It's free, please take it!				

Date				
Location				
Server				
Integrity: **this data is modified**	If Company Proprietary	Employee Private	Customer Private	**Business Partner, Government, Other**
The results will be absolutely disasterous with no chance of economic or politcal recovery.				
There will be severe financial, political or other undesirable results, but we will survive.				
It's gonna cost us big time, but spin doctoring will take care of it.				
Negligible effects, but we still really don't want it to happen.				
Publish it all you want. It's free, please take it!				

Partner, Government, Other: If you have information assets that do not neatly fit into the above three categories, stick it here. ☺

On the left side of the chart, I have arbitrarily picked five degrees of valuation and assessment.

- **Mission Critical:** If this info gets out, we are out of business. If this portion of our networks are attacked with DoS, we are out of business. If this hits the newspapers, it will cost us an absolute fortune and we won't recover from the losses. If this gets out, we go to jail. If this is attacked, the water supply (power, gas, other utility) is shut off. The Space Shuttle won't launch. The phones won't work. The bank stops functioning. You get the idea.

- **Almost Mission Critical:** In this case, the damage is tremendous, extremely costly (political, economic) but it is survivable. Barely. If these services go away we are going to be crucified by the media, customers will leave in droves, and we will go to the edge of disaster.

- **Ouch! This Really Hurts:** If this stuff gets out it will hurt; it will cost us, and it will sting, but we can take it, if we really, really have to. If this gets out, we can spin doctor our way back to health. If these services go away, we have a heck of a lot of catch-up to do and a lot of financial losses to contend with.

- **Negligible:** If this information is leaked or stolen, it's no disaster. It might cost us a bit; it might prove embarrassing, but there is no major impact upon the organizations' ability to do business. If this non-revenue producing web site goes down, that is not good, but it's just a disruption.

Date				
Location				
Server				
Availability: If these resources are destroyed or made unavailable	**Company Proprietary**	**Employee Private**	**Customer Private**	**Business Partner, Government, Other**
The results will be absolutely disasterous with no chance of economic or politcal recovery.				
There will be severe financial, political or other undesirable results, but we will survive.				
It's gonna cost us big time, but spin doctoring will take care of it.				
Negligible effects, but we still really don't want it to happen.				
Publish it all you want. It's free, please take it!				

- **Who cares?** Take the information, have a ball with it. We don't really care what you do with it. It's public, or we want it made public. If you modify it, no one will care or even notice. Denial of Service does not find a home in this category. DoS is bad, bad, bad – regardless of the form it takes.

The answers to "Who Cares" will vary depending upon which of the value charts you are developing. For example, you might not care that a specific set of information is released to the hoards; there is no inherent value in it to you or anyone else. (Confidentiality.) However, that same piece of information might create strong ripples (or worse) if it was modified, perhaps in some salacious way, to make an individual or company look bad. (Integrity.) You might not mind people reading the company phone book, but you don't want anyone changing the numbers. Access control rights generally include "Read Only" which is supposed to protect against data-integrity attacks. "Execute Only" should constrain integrity attacks against applications. Use what is available. Also, the inability to access that information or its associated resources might be more negative than the prior two cases. (Availability.) So, consider all three avenues in your data valuation process.

Part of the TBS security review process is to fill in charts like this. There are three steps to take:

1. Identify the existence of the information, which fits into these categories. This can be a very political exercise since so many people believe that their information is of high intrinsic value. So the auditors, beancounters, CFO and legal counsel should be part of the process to say, "yes, this is really, really critical," or "no, don't worry. It's of lower priority." Fill in one of these charts with some estimates from your perspective to get a feel for the process. Then, assemble a team to correlate policy against identified assets and fill in as many charts as you need. Break this task up among departments as well, to gain an intranet topological view of assets' locations and criticality.

2. Specify the logical, networked-based storage location of the assets you have categorized above. This process defines the pathing and logical organization of the assets with respect to the rest of the network.

3. Lastly, specify the physical location of the logical assets location in step #2 above.

By charting this process out, you will find a variety of conditions.

1. Mission Critical information may be distributed across several applications, running on different platforms and different Operating Systems.

2. There is only one Mission Critical application and asset base, and it all resides on one physical repository.

3. Who Cares information co-resides with Almost Mission Critical assets.

Understanding the matrix of asset location is critical to creating a secure environment, especially one that you can afford to build and maintain. By storing distributed Critical information all over hither and yon demonstrates poor controls, and makes security efforts all that much more difficult.

Using the Time Based Security approach, we do not want or expect all assets to be of the same value. We do not expect to secure an entire network or enterprise with the same diligence. That has been the Fortress Mentality approach for decades: secure the entire network as a singular whole, and treat all assets as equal.[31] The intent is to first identify the locations of critical assets, and then examine possible network reconfiguration prior to designing a security architecture that meets a company's real needs.

Consider you are faced with the following situation. By following these prior steps you have found that you have a myriad

---

[31] *The military addressed this exact issue with Multi-Level Security (MLS). They wanted to logically separate Confidential, Secret, Top Secret information using MAC, Mandatory Access Controls and 'object labeling.' Each MLS file is to be specifically labeled with a sensitivity flag. This way, a lower cleared user could 'write up' to a higher level, but not read the more sensitive files, while a more highly cleared user can 'read down' to lower sensitivity levels, but not write to them for fear of info-leakage. TBS does not deny MLS as an approach, but this first volume of Time Based Security does not address or model MLS issues.*

of sensitivity categories on a single computer; some are truly mission critical where you desire limited access to a small number of people, others are of inconsequential value but which are needed by a great many people on a regular basis. You are faced with a choice.

1. Design, implement and manage a security solution on this platform which serves both of your disparate security needs, or

2. Move one set of assets to a logically and physically isolated platform, and secure each to its real needs.

The first approach implies some sort of Multi-Level Security solution which is difficult to design and implement, not to mention terribly expensive to buy and maintain. The second is so much simpler and cost effective to accomplish. That reminds me of a story.

One of my financial clients said they were connecting to the Internet and were worried. I understood, or so I thought. I assumed it was their financial systems that concerned them, but they said, "no, it's the personnel files. Our lawyers tell us if any of them get out, we are in deep, deep trouble. What should we do?" This company has about 6,000 employees, and an HR department of about 14, all in one location. We looked at their topology, and then asked two questions:

1. Who needs access to the HR and personnel files? Answer: just the HR department. Even the CEO has to go through proper channels to access records.

2. What other company network resources does the personnel staff need? Answer: only local email and the Internet for Web and email.

The answer for them proved to very simple, inexpensive, and as a result, I got no immediate security work from them. "Put the HR records onto a sub-net that is only accessible by the HR staff locally. Provide no remote access to it, and add an air gap for security."

"What's an air gap?"

"The perfect firewall. There is no physical or logical connection, no network connection between the critical HR systems and the rest of the company."

"But what about email and the Internet?"

"Let HR keep their existing desktop computers for those functions. Give HR staff a second machine, perhaps only a Network Computer for the new HR net.[32] Absolute isolation, physical protection, and the only cost is a bit of hardware." Done. A low-tech answer to a high-tech problem that seemingly invited an expensive security product sale and integration effort. The air gap is the perfect firewall in many cases. This unconventional approach fits into the Time Based Security model perfectly.

With an airgap solution approach, from the company's enterprise network view we see,

F/BW = T,

and since there are no critical files electronically accessible from the enterprise network.

F = 0 Because the files' length is 0, therefore
0/BW = T and **T = 0** since **T = 0 = E = (D + R)**, no **D** or **R** are required on that subnet.

From the other side of the basic TBS formula, since the HR network is isolated with an airgap, the perfect firewall, from an electronic view, **P = ∞**, therefore there is no need for **D** or **R**. From the HR network view though, an independent TBS analysis is needed to determine what actions if any are appropriate. What do you think is the right TBS solutions approach for this company's HR network?

---

[32] *NC's are inherently more secure than PC's. Limited local resources, preferably no floppy or hard drive and no local ports. The object is PC functionality without the overhead of hardware, maintenance, upgrade and security. For those of you with gray hair like me, think "Smart Dumb Terminal."*

Whether you are practicing Time Based Security or not, the Data Valuation step is a truly critical one in an effort to defend yourself. It requires a level of effort, especially if it's never been done before, but I can assure you, some minor network reconfiguration and adding a computer or two, is one heck of a lot easier to manage and cheaper to buy than trying to secure an entire network as though everything on it is worth its figurative weight in gold. That false belief is arrogant, not true and can only end up hurting your overall security efforts.

With the Time Based Security formulas in tow, a data valuation approximation complete (maybe you have been able to fill in a chart or two from this chaplet), I am going to suggest that you ask yourself the following question: Based upon what you now know, do your networks really need any protection at all?

16
Do You Really Need Any Protection?

Of course, no concept like Time Based Security can go unchallenged and I always encourage debate and discussion. "Come on! Talk to me. What do you think? Does this Time Based thing-a-mabob make sense to you?" Slowly one question spurs another.

"What about the Coke formula?" One character from Pepsi asked. "Once you break into their computers, you can whisk it away in a microsecond. What could they do about that?"

"If the data is that totally crucial to the survival of the entire organization," I said to the Pepsi man, "and only one or two people need access to it ever, then it has absolutely no business being on a network. Networks are about communication and synergy, not isolation and segregation. Some things just don't belong on a network."

I stood and shouted. "Downgrade, m'boy! Downgrade. De-technify. Adding technology is not always the answer, and in many cases it is worse than the problem you started out with. If Coke has their formula on their network, go get it." The Pepsi man made a note of their URL. ☺

Sometimes, information has no business being on a network. The cost of protection, versus the risk, is just not worth it.

If $P = 0$ (where $D + R = E$), or
if $P < D + R$, and
F is a known quantity,

The files and information that can be stolen or irretrievably modified within the TBS environment are easily identified and known. Then the risk analysis and loss value is determined by the auditors and financial staff, not the security group.

As discussed earlier, sometimes disconnecting portions of a network makes a lot more economic and common sense than maintaining complex, perhaps untenable, or questionably secure interconnections. (No offense to any security vendor, but security mechanisms do add complexity and room for error.) Looking at network architecture and its functional topology is a useful security method than can avoid an awful lot of headaches, spending and administrative burden.

The second lesson that the 'Coke' exercise teaches some of us, is that our networked information isn't quite as important to our very existence as we thought it was. Maybe you want all of your information to get out, so there is no need for security at all. Or, maybe you only have one or two Mission Critical information components, and the best solution is to subnet them with reasonable protection, detection and reaction mechanisms. Applying the Time Based Security model to the way you conduct business will elucidate where you should best spend dollars and manpower in the safeguarding of your information.

A note, though, to some readers: If you are expecting absolute answers to your individual company's security problems, forget it. You won't get them here. First of all, Time Based Security is brand new, and who knows if the security community will adapt it. The prospects are good, but who knows. Secondly, as you've been hearing since the dawn of time, good security is requisitely predicated upon good policy. Translation: If you don't have a decent security policy, forget it. Forget trying to secure your networks or assets 'cause you really don't know what you're trying to do. Sorry for the brutal honesty here, but Time Based Security doesn't rewrite the fundamentals. It enhances them.

Lastly, before the security vendor community goes ballistic, I am not saying that Protection products are useless. I am not saying that vendors have been wrong or are building lousy products. Time Based Security is not meant to replace everything you have done. Do not automatically throw out your investment.

Time Based Security offers new tools to examine your security process, and as you shall see in a few minutes, how conventional Protection products can be enhanced with these new approaches.

17
Defense in Depth

In the hacker-ish movie *Sneakers*, the good-guys' goal was to enter a well-guarded room and steal the anti-crypto device. To accomplish that, a whole slew of defensive detection and reaction devices had to be circumvented. In other movies, the bad guys target a bank to rob it or they want to break into the rich guy's home to steal the Picasso or kidnap the children. Whatever. In each of these cases, though, an alarm system must be circumvented. That's part of the excitement of the cinema, and part of the real tasks with which bad guys must contend.

In one James Bond movie, the only way to reprogram the nuclear codes was with Presidential authorization which required retinal biometric confirmation by a staff officer. What did the bad guys do? Stole the officer's eyeball, removed the retina and surgically implanted it onto the bad guy's eyeball. He bypassed all security controls, stole the nukes and it was up to Bond to save the world.

In *Mission Impossible*, Tom Cruise worried that a single bead of sweat would trigger the detection/reaction systems while he hung there in a painful gravity-fighting sky-diving posture. In the movie *Executive Decision*, as in other terrorist/mad-bomber movies, the viewer experiences the same degree of frustration that the film's bomb squad feels as they encounter defense after defense. Defusing a well-constructed bomb is not so easy as cutting the red wire. Better bomb-makers put in detection circuits, which if clipped or cut, will trigger the bomb to explode. The design aim is to make defusing the bomb as difficult as possible by employing "Defense in Depth."

And the same concept applies to Time Based Security. A number of people have argued that "Time-Based PDR seems like a simple thing to get around. Why don't I just attack the **D**, the detection mechanism?"

In the case of the bank, why not just shoot the motion sensors and cameras? Or, why not cut the phone wires to the reaction troops – the alarm company – and there is no reaction? Let's examine the formula again where some unknown degree of protection, **P**, is assumed.

P > D + R

You know that if there is no Detection mechanism, $\mathbf{D} \Rightarrow \infty$, therefore, as a consequence, $\mathbf{P} \Rightarrow \infty$, and we are back to the old-fashioned Fortress Mentality model which we know can't work. Similarly, if there is no reaction mechanism, $\mathbf{R} \Rightarrow \infty$, as before, $\mathbf{P} \Rightarrow \infty$ and again we find ourselves in an unworkable situation. And this condition is what security has been based upon for years: no detection, no reaction, thereby providing the attacker with virtually unlimited time to launch attacks and the defender with no means of quantifying his defensive posture.

Assume that in a system analysis of a detection implementation, you determine that the biggest single weakness is indeed **D**. What do you do? In the older Fortress Mentality view, a manufacturer might redesign code or add additional barriers. With the TBS model, we take an integrated systems view and apply Defense in Depth by adding a new **P(d1)** layer to protect the first tier **D**. The TBS formula for this situations is:

P > D + R
$$\Downarrow$$
P (d1) > D (d1) + R (d1)

Just as in the case of additionally protecting a circuit in a bomb, **D** is deemed to require additional protection, which is then measured by the derivative formula, **P(d1) > D(d1) + R(d1)** where **d1** refers to the first tier **D**. In the case of our bank vault scenario, for example, the camera sensor might be the only detection device installed. But without it as a detection scheme, the sum of $\mathbf{D + R} \Rightarrow \infty$, unacceptably increasing the time value of **P** to cost prohibitive and ineffective levels. So, we choose instead to include a built-in protection mechanism to detect any tampering with **D**, or even, perhaps, an operational failure. So, now, for the system to work at minimum measurable effectiveness, the following conditions must be met.

P > D + R
P > 0
P(d1) > 0
P (d1) > D (d1) + R (d1)

This says that for the integrity of the first tier formula, **P > D + R** to hold throughout the system, the mechanism which protects **D** must detect and react attacks on **D** as if it were a stand-alone, independent function. For example, if **D** represents an electronic trip wire with a value of five seconds, we may choose to protect that **D** with a separate protection mechanism.

If **P(d1)** = 20 minutes, and
D (d1) + R (d1) < P(d1),

the derivative security does in fact protect the **D** mechanism in the first tier. Keep in mind, though, that there is no direct relationship between the behavior of the two isolated systems.

Of course, this logic can continue as deep as one feels is necessary to achieve the desired quantifiable level of security, as measured by time.

P > D + R
⇓
P (d1) > D (d1) + R (d1)
⇓
P (d2) > D (d2) + R (d2)

In the above example, because of the (hypothetical) sensitive nature of the contents to be protected, there are two additional tiers of defensive security to protect the first tier detection mechanism. Or, perhaps, budget constraints permit a layered approach where several inexpensive mechanisms are put into place as an alternative to much stronger mechanisms in a one or two tier approach. Again, if the simple mathematical methodology of time quantification is followed, the approach to adequate security is very straightforward, defensible, and measurable.

Assume now, that the detection scheme is determined to be adequate, but there are questions about the reaction mechanism. In *Surviving Denial of Service* chaplet, it is worst-case postulated that the primary means of control signal communications is blocked because of the nature of the attack. Therefore, a reaction along the primary communications media is not possible. Another path is required, ergo, another reaction mechanism is required.

$$P > D + R$$
$$\Downarrow$$
$$P\ (r1) > D\ (r1) + R\ (r1)$$

Here, the primary reaction path is arbitrarily considered to contain a weakness, worthy of further protection. If a bad guy attacks a system that he knows uses TBS, he might think of attacking the email or telephone Reaction paths before going after his primary target. He might want to interrupt the email to the administrator to increase the amount of unencumbered time he has available for his attack.

Security practitioners should not feel limited to these two examples. Combining additional layers of security is eminently acceptable. One might choose to enhance the system's overall effective security by further protecting both the Detection and Reaction components as in:

$$P >\ D +\ R$$
$$\Downarrow\quad \Downarrow$$
$$\Downarrow\quad P\ (r1) > D\ (r1) + R\ (r1)$$
$$P\ (d1) > D\ (d1) + R\ (d1)$$
$$\Downarrow$$
$$P\ (d2) > D\ (d2) + R\ (d2)$$

Here, two additional tiers of protection are provided for the detection component, and the reaction component receives one additional layer of protection. The strength of the first tier equation is then increased.

At the manufacturing level, instituting additional Protection mechanisms is nothing new with security products. Many high-end hardware-based cryptographic systems employ "tamper-proof" covers for critical components such as classified algorithms and key-management devices. While the algorithms and KM schemes are the tools to protect data, they themselves have often been considered the targeted items of value. If the "tamper-proof" cover was tampered with, the entire system essentially self-destructed, destroying any potential value to an attacker.

Time Based Security does not advocate placing le plastique inside of critical servers and network nodes, but the math and the model does aggressively suggest, and indeed insist, that a reaction is an essential element of defense, and must be taken as quickly as possible. Later in this book, you will come to see the implications of that approach.

Defense in Depth can take many faces, and in the next chaplet, you will see how it is applied in a sequential manner, and we evolve the concept of Key Node Defense.

18
SequentialTime-Based Security

When security is approached in the manner that TBS suggests, suddenly new views begin to come clear.

- All assets are not created equal, and they all do not deserve equal protection.
- Asset distribution by physical and logical separation is a security-process, but performed under the network architecture and topology banner.

Two approaches should be discussed when adding TBS to networks.

One implementation of Defense in Depth is sequential security, again borrowing from classical military-security thinking. Consider a secret military weapons storage facility. A physical one. First, the bad guys have to get onto the base without being captured or shot by climbing the electrified fences and shimmying through the razor wire. Then they have to invisibly meander to and break into the storage building, which will likely be surrounded by armed guards. Once inside, they would need to identify the proper storage location, and bypass that guard and sealed doors to get to the real target, the weapons.

In the real physical commercial world, sequential security, or Defense in Depth is also employed, but without the rigor. A high rise lobby has guards who will query your destination, but likely not shoot you if you herd onto an elevator without their permission. Locked office suite doors are the norm after 17:00 daily, and then the important folks lock their offices. Depending upon the contents of the office, additional security may be used in the form of locked file cabinets.

This physical-world view is easily applied within the enterprise, where multiple protective mechanisms can be used in large enterprises and within both Intranets and Extranets.

To take a simple example, let's assume that a company's Mission Critical data is contained within a popular manufacturer's database, on a large Unix server, buried somewhere within the corporate network. From the Internet's perspective, one must traverse through several mechanisms (whose specific functions are immaterial at this point) to get to the information.

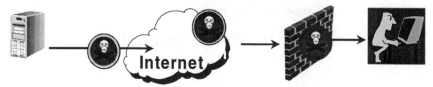

19
Key Node Security

The concept of Key Node Security is certainly nothing new. Large enterprises have long known that additional protection mechanisms are needed for higher sensitivity resources and assets. The use of internal firewalls and departmental Intranets echo that well founded philosophy: defend the most valuable assets with additional rigor.

Single Reaction Channel

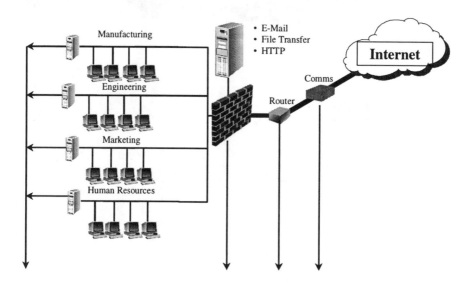

In the above network, let's assume that the really critical assets are in Engineering and Personnel. So, we added three Detection/Reaction mechanisms, one at the network perimeter, and two at the single access path to the resources to be protected. From the Internet, it certainly seems that there is sequential security in effect. However, that is misleading. Assuming no other external portals to this network, the outsider could attack Personnel by traversing through the perimeter firewall, heading over to the mainframe perhaps, and then migrating thoughout the network as an apparent insider. Also, the sequential security approach does not take into account the behavior of company and network insiders, which still account for a sizable proportion of bad-guy-like activities.[33]

In most networks, administrators and operations staff do not really understand the topology of their network, cannot absolutely confirm configuration integrity, nor know for certain that there are no PCAnywhere, modems or other external connection bypassing the network's security. And since we know that Fortress Mentality doesn't work, let's add TBS components at critical nodes as indicated above.

Using Key Node and Time Based Security, identifying anomalous or out-of-bounds insider behavior is an achievable goal. Placing Detection/Reaction components at critical or key nodes within a network is a cost-effective way to create additional defenses against insider threats. In large networks, then, we do not need to place TBS or Protection mechanisms everywhere; that is unnecessary overkill. Try this instead:

- Identify the critical assets
- Place the critical assets in appropriate physical and logical locations
- Add TBS protection at the critical nodes.

However, to maximize internal (as well as external) defenses, we have to expand our concept of what we mean by system monitoring. More about that in the next Chaplet.

The data base server sits within an intranet, which in most organizations suggests an additional layer of security, as noted. The application itself might contain some additional access control mechanism, and the host server might employ detection systems of some sort. Passwords might be used, firewalls might be employed. But, ultimately what we care about is the detection and reaction systems that are placed in the route from the Internet to the data

---

[33] *Studies show that between 40-80% of known computer crime occurs either by, or with the help of insiders. Reference studies by the Computer Security Institute, the FBI, ICSA, Warroom Research are at their respective web sites, or linked through www.infowar.com.*

fields in the database. The simplest way to view the effects of Time Based Security is to view each implementation as independent, with no interactive effects.[34]

Sequential Time Based Security

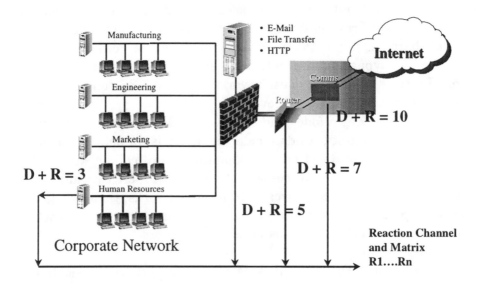

Thus, a password nexus would result in a given projected value for **P(pw)**. A TBS mechanism at a firewall nexus (say at the Intranet) would result in a separate **P(fw)** and the operating system might have a **P(os)**. Under the (weak) topological assumption[35] that there is no network alternative in routing from the Internet to the target data files, (an administrator's determination), then we have a data protection value in TBS terms as:

$$P\ (db) = P\ (pw) + P\ (fw) + P\ (os)$$

---

[34] *In more sophisticated security environmental, an integrated TBS implementation might involve a Detection and/or Reaction Channel where the **D** and **R** components are tied together. More about that later.*

[35] *In the real world, networks are not linear. There are multiple paths from A to B. Sequential TBS is a conceptual stepping stone to Key Node TBS, coming up shortly.*

This simply says that the intruder must first conquer the password mechanism at the periphery of the system, then the Intranet firewall mechanism and finally the OS or application security at the host. If we apply TBS to this picture, we see that the defenses expand to:

$$P\ (data) = P\ (pw) + P\ (fw) + P\ (os)$$
$$\Downarrow \qquad\qquad \Downarrow \qquad\qquad \Downarrow$$
$$\Downarrow \qquad\qquad \Downarrow \qquad\qquad P\ (os) > D\ (os) + R\ (os)$$
$$\Downarrow \qquad\qquad P\ (fw) > D\ (fw) + R\ (fw)$$
$$P\ (pw) > D\ (pw) + R\ (pw)$$

In this example, each **P** component has its own **D** and **R**, working in isolation from the others. So, at **P(pw)**, a detected anomaly will create a reaction, if properly done. Administration will know that someone is attempting to, or has broken through the network perimeter. But, they have no information on what their real target might be. The only thing that can be said is, "the network is under attack." [36]

Note that in Sequential TBS, the presumed time values of each **P** add together for an estimated total **P(data)**, but that estimated amount of protection time is only applicable to the contents of the data base, not the other information assets of the network. The obvious disadvantage is that such sequential and additive values for **P** are only effective in a very few cases where the network topology and pathing is severely constrained. Key Node security will solve this problem.

---

[36] *However, performance can be enhanced, and cost perhaps reduced by integrating the **D** and **R** functions from one tier to the next. (See the chaplet on Reaction Channels.)*

20
Reaction Channels

Let's head back to the physical world
for a moment, and consider a large corporate campus. Perhaps
there are a dozen or more buildings, hundreds of offices, or maybe
multiple floors within a high rise. At the elevator doors, building or
floor entrances, individual office doors, windows and perhaps
maintenance portals, the security acute organization will install
sensors (detection). If a worker enters the building or office, he has
to enter the right code in the alarm system or take some other
action, which shuts down the reaction mechanism. But in each and
every case, the reaction mechanism is probably going to be the
same: call the police, or alarm company cops or someone else in
authority whose goal it is to thwart the intrusion.

And why should it be any different in our network
application of Time Based Security. The answer is, it doesn't have
to be any different.

Single Reaction Channel

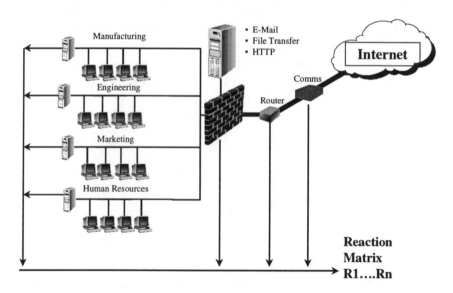

In the above network, a number of Key Nodes have had TBS mechanisms installed. Each of the detection mechanisms in this case operate independently of each other, which is just fine since each environment is slightly different, and the relative sensitivity of the protected resources varies.

However, in some cases, it makes a whole lot of sense to develop a common means of Reaction. This network uses four TBS detection mechanisms which each drive a common reaction mechanism. In the above network, a number of Key Nodes have had TBS mechanisms installed. Each of the detection mechanisms in this case operates independently of each other. If we can momentarily assume a common protocol, the Reaction channel communicates: (for example)

- the source of the detected anomaly,
- the date/time stamp of the event
- a log of the anomaly
- a sensitivity label [37]

We want the Reaction Channel to communicate as much relevant information as it can, so that an appropriate reaction can be taken. Further, the more information available at that point will increase the speed of the reaction, thus decreasing the value of **R**, which is one of the goals of TBS.

In the case of a single reaction channel, the reaction device at the end of the chain must be able to rapidly decipher the transmitted information and take a reaction. If the end of the single reaction channel is a human being being notified of a detected event he needs concise information to make a rapid, well-informed decision. He could be notified by telephone where a voice interprets the reaction channel protocol and data, speaking in English to the administrator. If email to a human is the chosen method, then the

---

[37] *This is the beginning of establishing a reaction protocol for system interoperability and compatibility. In the coming months, we hope to assist in the establishment of a TBS Consortium, where such protocols and common criteria can be agreed upon.*

format of the data supplied by the detection mechanisms is critical to rapid interpretation and decision making.

One of the protocol components which will assist in the speed of response and reaction is a "Recommended Reactions" based upon policy. (Reaction Matrices are covered in the next chaplet.)

Regardless, though, if there is human intervention in the reaction process, the value of **R** will increase, as we showed in the example with Administrator Dick quite a while ago. In TBS parlance, this is not good. ☺ If the policy is clear, and the organization is willing, creating an automatic reaction based upon detected events is the best mechanism to deliver strong protection through high-speed detection and reaction. (Don't forget the basics: **P > D + R**. We cannot assume that protective devices alone will deliver the level of defense we really want.)

Another way to implement TBS Reactions is through multiple reaction channels, which might be preferable in some situations.

Multiple Reaction Channels

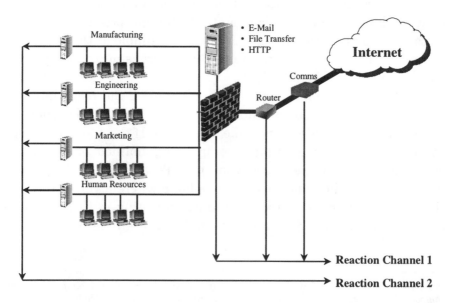

In this network example, the perimeter TBS detection systems drive Reaction Channel #1, which results in one set of pre-defined reactions based upon detected events and policy. The Key Node TBS detection mechanisms at the subnets, though, drive Reaction Channel #2, which would generate a different suite of reactions, perhaps more automatic, perhaps more severe. This is one of the obvious applications of Data Valuation and data separation.

Of course, each and every TBS node can have its own Reaction Channel, but the complication of the administration and management of the security infrastructure might outweigh the benefits. As we'll see in the next chaplet, a well thought out set of Reaction Matrices is the most propitious route to take in enterprise-wide TBS implementations.

21
TBS Reaction Matrices and Empowerment

We know how to handle the Acts of God. It's the Acts of Man we haven't yet faced square on.

It should be inherently obvious at this point, (unless I have utterly failed in my task) that in better Time Based Security implementations, the value of **R** should be as small as possible (as should **D**). As in the previous chaplet, we assume that our detection systems are AOK, and the next TBS step is to trigger a response of some sort.

Network

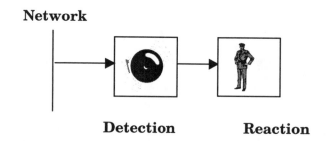

Detection   **Reaction**

Whether it's armed guards arriving at the scene of the vault robbery within 180 seconds of being alerted, or an automated software response reacting in less than 1 second, a lower value for **R** will reduce the necessity for reliance upon a more comprehensive **P**. A lower **R** also reduces the potential costs associated with erecting and managing a really strong **P**. A very rapid automated response in many cases is the most expedient choice, especially if it is protecting tenuously guarded electronic assets.

Determining the value of **R**, though, is a 3-step process, as we saw with Administrator Dick earlier. The first component of Reaction is Notification which communicates to someone or something, that the detection system has been triggered.

With many network-based automated detection systems in use today, once an anomaly is detected, the reaction is to merely notify "someone in charge" so they can decide what to do. Notification of "someone in charge" generally means "notification of someone other than the boss who doesn't want to be bothered at 3:00AM," which serves only to increase the value of **R** and decrease the effective security of the network. In human-intervention reaction systems, notification can take several courses. Each needs to be evaluated, and perhaps fine-tuned, to keep that **R** low.

A first step in decreasing **R** is to fill out a notification chart, and measure how long these varying reaction processes take. (Of course, you can modify this chart to fit your own needs.) By now you should have a target value for **E, T** or **R**, and that value can be entered into Column 1. Then, based upon the nature of the notification system and process choice, enter the predicted value you hope to find in Column 2. Measure the system and enter that answer in Column 3. Documenting this exercise is important for your own security evaluation process, but perhaps even more importantly, it is for management, so that they can understand the quantitative nature of Time Based Security. (This helps you get more budget dollars for your job! In a sort of *Dilbert-ish* way, more paperwork is good.)

Second, in non-automatic responses, determine how long it takes to get to a place where you can actually fix the problem.

The third step in the Reaction process is to actually repair the problem. What is the actual reaction to take to resolve the situation, and how long does that process take? It is at this point that the company must invest some time and energy to translate corporate policy into TBS-Reaction Matrices. A Reaction Matrix should ideally consist of pre-thought out reactions to given sets of conditions or events that have been detected. Look at the Reaction Matrix with a few examples; again, we are concerned with the amount of time it takes to actually deal with the detected anomaly.

This is a very brief example of what an organization needs to do in developing a Reaction Matrix: consider the type of attacks it might encounter and therefore detect, whether they are external sniffing, internal host-based audit trails, or unacceptable network behaviors. The Reaction Matrix development should be based upon policy, contingency management and the amount of risk the company is willing to take – ultimately measured in terms of Time, as in Exposure Time, **E**. This process should be nothing new to those of you familiar with disaster recovery (et al) procedures.

Notification Means (Reaction)	Desired Time	Predicted Time	Measured Time
During Work Hours			
E-Mail to desk (At peak Traffic Times)			
E-Mail to desk (At off-hours)			
E-Mail when not at desk			
Pager with return phone # or 911			
Page with full message			
Phone call to desk			
Notify 2nd in charge (admin)			
Not During Work Hours			
E-Mail to home (At peak Traffic Times)			
E-Mail to home (At off-hours)			
E-Mail when not at home			
Pager with return phone # or 911			
Page with full message			
Phone call to home			

In the well-budgeted security world of disaster recovery, companies spend a great deal of time and money planning for the Acts of God. How can we keep the transaction processors operational when the hurricane flattens the data center? How long does it take to get operational after the highway overpass tumbles on top of our building in the midst of a 7.4 earthquake? When downtown Chicago basements flood, how can the call center service its customers?

How can Home Shopping Network keep selling parasols to its California clientelle, when its Florida broadcast facility has been struck by lightening 300 times in the past hour?

And so it goes for tornadoes, power outages, trees collapsing on communications lines, solar flares... and any other Denial of Service events we can blame on Mother Nature. There are exceptions, of course. When the World Trade Center was bombed February 26, 1993, many of the businesses had arrangements for off-site recovery and continued operation with minimal (relative) disruption. Those without such contingency plans were in trouble.

One report said that 23% of businesses in the World Trade Center were out of business within one year.

However, in March of that same year, the *No-name Storm*, or *Storm of the Century* battered the East Coast, and a tremendous amount of snow fell on Long Island, New York. The roof of a banking center which controlled seven ATM networks collapsed under the strain, stranding nearly 30 million ATM customers for up to one month. Problem? The off-site recovery centers were already full from the bombing. The double whammy had struck; a statistical improbability, and a set of conditions that no one seriously considered.

Reaction Matrix			
		Desired	Measured
Detected Event (Anomaly)	Chosen Reaction	Time	Time
3 Bad Password Attempts	Log and Notify Admin	1 sec	2.4 secs
3 Bad Password Attempts	Turn off Account/Notify Admin	1 sec	.94 secs
Mulitple Port Scan	Initiate Trace Route	250ms	1.5 secs
Internal User - Audit Bahavior #1	Involve HR Immediately		
Ping of Death	Kill the Bastard :-)		
Syn-Ack Attack	Reaction # 23		
Mail Bombs	Reaction # 81		
Firewall Breach Attempt	Autofilter Source	100ms	2.7 secs
Traffic 2X Anticipated	Log and Notify Admin		
Multiple Site Attack	Shut Down Network	3 secs	2 Days
Shut Down $ Server	Isolate Network	1 min	2.4 hours

Time Based Security says we should consider the Acts of Man as diligently as we do the Acts of God. Through the use of Reaction Matrices, an organization should develop a set of contingency plans (Reactions) that are consistent with the detected events and risk potential for loss.

In the Reaction Matrix above, an administrator is the human involved in the response chain and also represents the step with the greatest room for error[38] because it increases **R** unacceptably. Most people hesitate to make the tough decisions quickly. Thus, especially in those cases where **P = 0** (or we assume

---

[38] *Remember the Government security manager from earlier, whose idea of security and response was to manually examine his audit logs every couple of weeks.*

it to) and the goals it to limit **E**, a human reacting to solve a problem in ten seconds by punching the correct key on the administration keyboard can be entirely too long.

To increase the defense of the network, an automated response mechanism is certainly a choice that Time Based Security encourages, but this is only viable if the appropriate responses are predetermined based upon a set of input conditions. Thus, if a Denial of Service attack is under way, a pre-described set of responses should be considered, and programmed into the response suite. However, if automatic reaction is chosen, one coincident response should be to contact the administrator on call and advise him of the attack (or other security violation) and the actions that were taken automatically without human intervention.

The Administrator will know exactly what the system did to protect itself, and what state it is in the moment he takes over. The Administrator can then, with greater leisure, manually invoke other reactions, shut down the automated responses or even try to strike back at the offender.[39] If implemented properly, **R** remains small, and the human element does not degradate **R** or the defensive process. Creating an automatic Response Matrix is critical to efficient TBS as it also enforces policy. But, even more importantly, the Administrator must be similarly empowered to make tough decisions.

Assume that based upon a certain detected event, the Reaction Matrix says to disconnect a subnet of the organization's infrastructure from the rest of the enterprise in order to protect its assets. The people within that division might not exactly be happy about being disconnected from the world, but that particular reaction was chosen to isolate the attack and limit damage. The so-called sacrificing of a pawn to protect the king is a well-understood and rational strategy in any form of conflict – and applies in network security as well.

---

[39] *An issue to be considered is whether automatic responses should tend towards the false-positive or false-negative. A false positive detection/reaction could unintentionally interfere with normal business operations which is not on the highly-desirable list. But false-negatives can permit certain events to occur without a defensive response. Policy and risk decisions must be the deciding factors.*

Further, such pre-programmed reactions can insure a graceful degradation of a network, system, process or service when appropriate. Shutting the whole 'thing' down, although perhaps very painful, is far superior to what would happen if it stayed up and under attack. When some of the more significant automatic reactions occur, the Administrator is called by the parallel reaction process, and then he is faced with answering some tough questions and perhaps making some tough choices:

- Is the attack real?
- What was the goal of the attack?
- Is the attack still occurring?
- Did the Reaction Matrix come to the proper conclusion and act accordingly?
- Was the attacked thwarted?
- Do we need to perform a damage assessment?
- What further steps are needed to protect the organization and its resources?
- When can he/she get the systems back on-line?
- Is further disconnect and systemic shutdown in order?
- Lastly, who is behind the attack and can we nail him/her?

Administrators must, now more than ever, be empowered to make these sorts of tough decisions that affect business operations. They must be empowered by the top management in the organization – and supported when they act according to policy. Security administrators are historically placed in an unenviable position: they are held accountable when things go wrong, but not given the authority to take drastic (re)actions without the blessing of the IT/IS/MIS manager or Board of Directors. Any such additional human intervention in the Reaction process is anathema to effective Time Based Security. Empowerment by management to the front-line network administrator is necessary to limit R, and effectively increase the defensive posture of the network. It may be a tough pill to swallow for old-time managers, but a necessary component of TBS.

Upgrading Detection Through Reaction

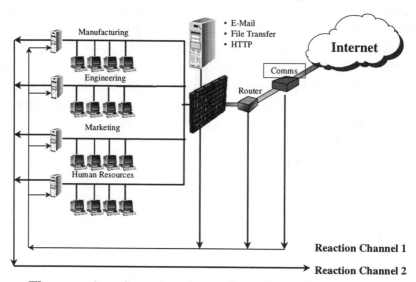

The security of a network can be enhanced by implementing an iteritive, or feedback-based process in the Time Based Security detection/reaction mechanisms. In the example above, the host based detection mechanisms (in this case, **D(os)** at the departmental servers) receive advance notice of a perimeter intrusion (or attempt) via Reaction Channel 1 from the detected events at the router and firewall. This type of advance notice to downstream (inner network) detection mechanisms could cause them to operate with a greater degree of sensitivity than they might in 'normal mode.' The detection mechanism, though, must be able to be remotely reconfigured to meet such system requests. (Attention vendors!)

Conceptually, the feedback and integration of the detection and reaction mechanisms creates a security architecture with a holistic enterprise view, which increases efficiency and strengthens defensive posture. While there is no way to specifically know that an assault at the perimeter of a network implies an assault upon a specific data base or resource, advance network notification, upgrading detection sensitivity for some period of time and increased awareness is a sizable benefit over today's security methods.

22
Gaming & Simulations

This chaplet is a short conceptual rant, so bear with me. Or as Garfield might say, "Get over it."

As I have hopefully reinforced in the last two chaplets, part of any network defensive process is developing the big *'what if?'* questions that we generally do not allow ourselves the luxury to ask and answer. Budgets are tight, and time is tighter. Who the heck has time to sit around and ask *'what if'* questions anyway?

The questions have to be both asked and answered. In a chess game, often the winner is the player who can look ahead more moves than his opponent, predict his moves and choose a better strategy. It also helps when the other player makes a mistake.

A well planned military campaign is also based upon a long series of *'what if?'* considerations. During the Cold War, think tanks like Rand and Hudson were paid a great deal of money to figure out how nuclear wars would swing, in our favor or the Soviets', depending upon a long and continuously changing set of political, economic and social conditions. And then they would play the games all over again so that the political and military leadership of this country would theoretically have the best decision-making information at their fingertips. If this happens, only 100 million Americans die... if that happens, then 150 million will die. Our *'what if'* postulates fell smack-dab on their derriere, though, when our intelligence community failed to predict the democratic uprisings in Russia and the India/Pakistan nuclear arms testing to name two. (As a network administrator, I hope you will develop better *'what if'* predictions than they did.)

After the publication of *"Information Warfare"* in 1994, though, the concept of battle without bombs, bullets or bayonets was forced out of the classified world and into the lexicon of both network defense and national infrastructure protection. So, in gaming scenarios like *The Day After* (1995-6) and *Eligible Receiver* (1997), contingencies were examined to provide the players with some handle on how to respond to real-life emergencies.

I had the honor to head up the Cyber-Manhattan Project's Electronic Civil Defense Team where our charter was to develop scenario and role-playing games to create realistic training. Fred Villella, former senior FEMA executive[40] and I have conducted live exercises for national, state, local and corporate management where both realistic and impossible situations are staged as a training aid. This work is very fulfilling, but such efforts must be brought closer to the network administration level.

Thus, I am a big fan of games.

I am a big fan of engaging people in spirited interactive discussions or activities. In very intense bursts of role-playing or gaming exercises, immense learning can, and does take place. Just like kids learn better when they are enjoying themselves, seasoned professional adults learn better when the *'boring barrier'* is crossed and the brain is actually receptive to new information and ideas. Besides, like much of the planet's population, I just like games.

There are different kinds of games. Solo (solataire-ish games), 2-person strategy (Chess, Stratego), 2-person strategy+luck (Backgammon, Card Games), multiple person board games, role-playing and so on. Lots of ways to play games, and in the last couple of years, I have had the opportunity to work with lots of different folks, all of whom have different spins on gaming. In many cases, the intent has been to use gaming and role-playing exercises as an educational tool for information security issues, policy development, Infowar scenarios, contingency management, situational response, government interaction, and related interests.

Gaming is effective at several levels within organizations. For awareness and general security training, large numbers groups of people can be put through a series of exercises which puts them under stress and faced with intense time-critical crises. Non-technical management can be put to the test in higher-stress exercises, often faced with highly technical decisions. We have run exercises for municipal governments and emergency services, which address enterprise-wide and interenterprise or infrastructural

---

[40] *Federal Emergency Management Agency, which was responsible for the media-bashed Continuity of Government planing.*

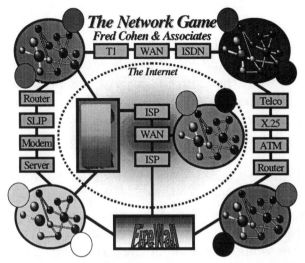

The Network Game
Fred Cohen & Associates

dilemmas. When done properly, participants are operating (playing) at peak efficiency, and enjoying themselves.

The reason for this minor rant on why I love games (and so should you!) is because to maximize the effectiveness of Time Based Security, the Reaction Matrices really need to be seriously considered. I recommend to many different organizations who are looking for higher levels of security, that role-playing and gaming be used as techniques to examine situational problems and develop planned responses.

As your reaction matrices show, and indeed what this whole book is about, time is of the essence in protecting information assets. Therefore, faster reaction matrices are good things, and some well reasoned, pre-planned responses can add to your overall security efforts.

Of course, we end up back with policy and the early games in any company should begin at the highest levels with the President, CFO, COO and other significant decisions makers.[41] You gotta isolate these guys for a couple of days, and really get them to understand the consequences of their choices (or procrastination), and to hopefully realize how absolutely reliant upon the technology we all are. In these sorts of interactive exercises, potential network attack/failure scenarios are postulated and then appropriate responses are worked, developed, documented and used as guidelines or automated into the reaction process. Thus, for security to be truly efficacious, management must think about the constantly changing threats and vulnerabilities, and they must

---

[41] *The Network Game was developed by Dr. Fred Cohen. Details can be found at all.net. The TBS Game is at www.infowar.com.*

think about and choose their first responses. This sort of exercise wants to be conducted on a periodic basis.

Options, alternatives and additional fail-safe procedures are part and parcel of any well thought out security effort, but especially TBS. Remember, time is the critical element, and by reducing all components to a time-based measurement, we find a common denominator by which a reasonable comparison can be established.

Graceful Degradation should also be part of any serious consequence management exercises. It begins with, "if all heck breaks loose, how can we stay in business?" We are all so absolutely reliant upon technology for our day to day existence, and we cannot fathom how our businesses would run without the technology... but this is just the process I am suggesting.

Graceful Degradation means that if IT hits the fan, an organization has developed a process by which it can survive and operate on some minimal level without its complete technological foundation. In some cases, an organization can revert to paper handling, just like the good ol' days; but this sort of performance and operational degradation only works if it is well-planned. Otherwise, if the attack or event is severe enough, the fallback position is anything but graceful.

Many organizations have installed some forms of redundant technology in the event major system components fail, and that is good. But what about technology-less redundancy? How many organizations have gone through the process of figuring out how to function without computers or fax machines or email?[42] The malicious Acts of Man are more likely to be a more thoroughly disabling attack than any hurricane.

I whole-heartedly encourage developing a Graceful Degradation plan as part of any serious security efforts. So, please do consider gaming as a means of education, awareness and training throughout the entire organization and to engage senior management in policy and reaction development. And, as usual, let us know how it goes and if we can help. End of rant.

---

[42] *The results of this exercise are not pleasant, but training and contingency management makes an organization more able to deal with significant systemic disruptions.*

23
Using TBS in Protection

So far, we have only looked at Time Based Security as a Detection/Reaction mechanism, relegating Protection to the sidelines. The use of TBS, in an effort to merge components of risk analysis and classic info-sec, often assumes $P = 0$, which again says protection is nonexistent and/or useless.

That perception is entirely wrong. Well-managed, properly configured protective mechanisms are a good thing. They enhance security, but for all of the reasons given earlier, we cannot completely rely on protection to defend our assets and resources. No rehash necessary. So, let's examine how Time Based Security can enhance the protection element of our basic formula, $P > D + R$.

In a generic protection process (such as passwords, access control, etc.), the user will cause a process request to be initiated. On the Internet, for example, if you request a protected URL, a

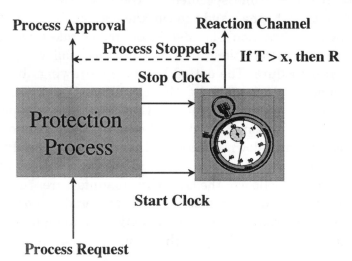

password script is invoked. In a Novell or NT server, access control tables are examined based upon identity and time/date to determine your rights. These Fortress Mentality protective devices can be strengthened by adding TBS mechanisms as above.

As soon as the protective process is initiated, a "Start Clock" signal should be triggered, just as one starts a stop-watch for timing sports events. The clock initiation should occur as quickly as possible (within microseconds). The clock itself is programmed to run for 'X' period of time, where **X = The Maximum Amount of Time For The Protective Process To Function**. Said another way, each protective process takes some reasonable amount of time to be completed, and we want to define the outer limit of that time, 'X'.

The clock continues to run until the protective process sends the clock a subsequent Stop-Clock signal, which means, "this protective process functioned correctly, within the specified time limit established by 'X'. In these instances, the Clock circuit becomes a Detection mechanism. If the Clock/Detection circuits receive the "Stop Clock" signal, the Clock-Detection circuit sends no signals to the Reaction Channel. Everything is working properly.

If, however, the protective mechanism is kept open for longer than the 'X' time specified by the clock circuit, it will not send a Stop Clock signal, and the clock/detection circuit will send a signal down the Reaction Channel. Then, policy takes over and the results of the Detection/Reaction process follow previously determined procedure. The dotted line in the drawing above shows that in some implementations, the clock/detection mechanism is tightly interwoven with the protection process. Thus, the protective mechanism manufacturer could 'hard-wire' its own reaction, and if **T > X**, could shut down access to the output of the Protective Process channel.

It would behoove the security manufacturer or security integrator who builds TBS mechanisms into products or networks to provide an external driver to the Reaction Channel in addition to integrated any self-protecting mechanism.[43]

---

[43] *A vendor agreement on Detection / Reaction protocols will propel these concepts into reality.*

Time-Based Security in I&A and Access Control

Some Identification and Authentication (I&A) methods inherently employ the TBS concepts, but not with the formality suggested here. The office worker who must run to the closet and affirm his/her legitimacy by entering an appropriate ID code into the alarm is an easy-to-remember reference model.

One way in which the TBS model is applied for I&A, is to first define the **P,** as the protective window of time defined by the manufacturer (and/or configurable by the user) as "the maximum amount of time one permits to transpire to accomplish proper authentication," as in the I&A drawing. The vendor should include a configurable option where you choose what amount of time, **P,** is expected for the UserID and Password to be entered and authenticated. Within 3 seconds? 10 seconds? 30 seconds?

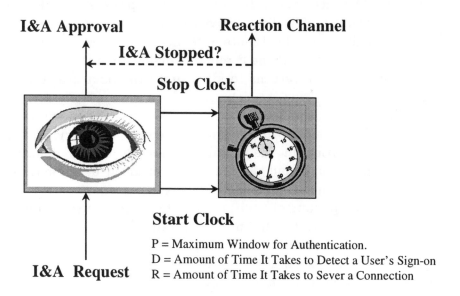

The value for **D** which is how long a good authentication actually takes, is some amount less than **P,** and of course **R** represents how long it takes for a chosen reaction to occur. **P,** with its predetermined (and/or configurable) fudge factor, must be more than the sum of **D + R**.

When a user first contacts an I&A mechanism, the detection process begins. Part of that detection process initiation is "waking-up" the response mechanism for the fastest possible reaction when it's needed. If, for any reason at all, the authentication process does not occur within the specified **D**, then a reaction is triggered such as severing that user's connection in **R** (as one possible reaction.) I&A logs should be kept, and if the I&A process was also violated in other ways, (i.e., three failed attempts) then further policy driven, non-real time restrictions can be put into place.

Once the user is authenticated in the normal manner as prescribed by the I&A product, in a time period less than **D**, an optional small amount of time delay can be added to the process prior to passing the user through the protective security mechanism. The rationale is that the detection process is useless if it takes more time to detect the activity than it does for the activity to occur. This delay in a legitimate authentication process, which is faster than **D**, insures that the **P** remains higher than the sum of **D + R**. The system then returns to idle – until the next user comes along.

Thus;

P = Effective security provided, measured in time, for a proper authentication in the usual manner plus some nominal delay. (In well thought out designs, the embedded delay may be variable as detection and reaction processes improve.)

D = The amount of time it takes to detect a user's initial sign on and determine its authenticity or fraudulence,

R = The amount of time it takes for the detection module to trigger the reaction, such as severing of the open port in question, or any other chosen reaction process.

Time-Based PDR & Access Control

Access control mechanisms conventionally provide a fortress mentality barrier to entry into computer systems or physical locations. The historical premise has been that a Reference Monitor, rules-based mediator provides sufficient defensive protection in computer systems. The development of such systems that meet stringent controls is a lengthy, expensive process, and somewhat outdated. Using Time Based Security, the design and management constraints of access control mechanisms can be relaxed somewhat, in favor of enhanced detection-reaction schemes.

P = The maximum amount of time it takes to provide legitimate access to resources as determined by the Access Control Tables or other mediated access process added to some nominal delay introduced to maintain the integrity of the TBS formula. So, **P = P(native) + Delay**

D = The amount of time it takes to detect a user's initial request for access and determine that the request was either in or out of bounds.

R = The amount of time it takes for the detection module to trigger the severing the connection, or deny all access to that user or class of users or invoke another reaction process.

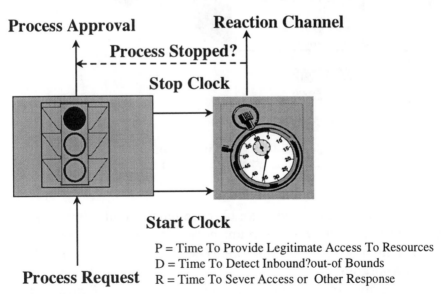

Process Approval   **Reaction Channel**

Process Stopped?

Stop Clock

Start Clock

Process Request   P = Time To Provide Legitimate Access To Resources
D = Time To Detect Inbound?out-of Bounds
R = Time To Sever Access or Other Response

Thus, as in a prior I&A example, a slight delay (nominal, and somewhat arbitrarily based upon the evaluation of **E** and **T**) will be forced at **P** to maintain the integrity of the fundamental TBS equation.

Time-Based Security & Enterprise Audit Trails

In an enterprise, disparate computing systems often provide an auditing capability to record user activities, administrative actions and security relevant changes. All too often, though, the

Enterprise Audit Trails and TBS

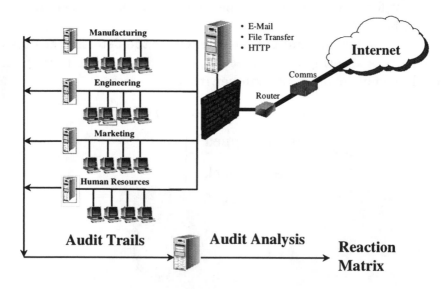

audit trails are relegated to an obscure partition of a hard disk and then perhaps read by a human, once every so often, as with our Government security officer. Clearly, with the TBS approach, this is unacceptable.

A far superior audit trail architecture in a distributed TBS environment is a real-time monitor at every Key Node (critical nexus or server), reporting to a central point where analysis is performed, as in the drawing below. The audit trails created at each key Node are themselves not the detection system according to TBS. They are the monitors that collect distributed data of interest, then feed it to a centralized "audit analysis server". Based upon the policy-based interpretation of the raw data at the audit analysis server, the results can trigger a response of some sort.

In this case, we have no idea if there is any protection mechanism at all, or any series of protection mechanisms. If it's a typically large network, there are long forgotten interconnections, so it is prudent to assume we are dealing with a $P = 0$. In the audit trail applications of TBS,[44]

$P = 0$

D = The amount of time it takes a platform specific audit device or monitor to record selected behavior, transmit it to the central detection processor, analyze the data and come to a conclusion.

R = The amount of time it takes for the detection module to trigger the reaction process, and for the chosen reaction to take place.

Time-Based Security & Encryption

Encryption in its many incarnations is touted as a powerful, effective, inexpensive and easy-to-use means to provide information with confidentiality and integrity. Some encryption products are invisible to the user and work quite well at their designated tasks. Others are so cumbersome, it's far easier to siphon vodka from a cow's udder.

The fundamental problem is an inherent misunderstanding as to what really constitutes security in an encryption system. The popular argument we hear is, "longer key length is essential for good security. Shorter key length is an invitation to criminals and

---

[44] *In the chaplet "Other Detection Means", we will see other similar Detection/Reaction schema that do not have specific integrated protection mechanisms either.*

government abuse." Well, sort of, and Time Based Security gives us a clue as to what really matters. Encryption utilizes two defensive mechanisms to protect information.

- The encryption algorithm itself, which in better systems is an open published technique. The amount of time required by an intensive cryptanalytic or brute force attack is the measure of the algorithm's seeming strength.
- Encryption keys which come in lengths from 40 bit (very weak) to 2,048 or more (very strong), and they must be given additional protection to defend the contents of the encrypted information.

With Time Based Security, encryption is viewed as a technique to:

- Delay the disclosure of the private contents of a message (Confidentiality), or
- To permit detection of an otherwise undetected alteration of a message. (Integrity.)

With respect to confidentiality, the effectiveness of encryption in TBS terms is measured by:

P = The amount of time required by a brute force (or cryptanalytic) attack with no plaintext samples available, for a given encrypted message to be decrypyted. This value can be somewhat arbitrary since exact decryption knowledge is also deemed secret by many government organizations, but is meant to serve as a guide. In a brute force attack, the total number of encryption key possibilities divided by two, divided by attempts per time unit gives us the average time to decrypt.[45]

D = The amount of time that the protected information has any real and/or perceived value.

R = There is no reaction component for the encryption application of the TBS Model.. Thus, in this case, the formula is simply reduced to

---

[45] *In July 1998, Deep Crack, a parallel processing computer (MPU) solved a single level DES equation (56 bit) in 25% of the total brute force time, or 50% of the average predicted time.*

P > D + R, where **R = 0**

The popular conception that key length is the most critical element begins to waver when we look at the case of on-line commerce. Encrypted tokens (or passwords or session ID's) for on-line commerce are only used for a few milliseconds and generally for a single session transaction.

If **D = 200ms**, a reasonable time for Internet commerce, then an encryption scheme which offers any greater level of security could be considered adequate to hold the basic TBS formula true, given that the key management method was sufficient. So, even 40 bit or 56 bit (exportable) encryption methods would be more than sufficient if in fact the information in transit was only valuable for a few milliseconds.[46]

However, the strength of cryptographic systems cannot be measured solely upon key length and algorithmic integrity. In public key infrastructures (including Certificate Authorities) the data bases which hold the keys and associated security relevant information is a greater potential system weakness – as well as a potential single point of failure. Since the cryptographic keys to the system are an obvious target, they must be protected with a degree of vigor; **P(km) > D(km) + R(km)**, (km = key management) which echoes other conventional TBS approaches of information defense.

On the other end of the spectrum, such as in the case of trade secrets or other proprietary information, the information could have long term value so **D** could be years. In that case, both **P** must be very high, and the **P(km)** must be correspondingly high. Unfortunately, maintaining a high value for **P(km)** is a far more difficult exercise than picking a thousand bit key.

In the next chaplet you will learn that there are a lot more commonly available tools at your disposal (and some not yet available) to further enhance the effectiveness of Time Based Security and the defense of your networks and information assets.

---

[46] *Not really; there are even more considerations. See the Chaplet "Other Detection Methods"*

24
More Detection, More!

I remember my uncle, and he was a jerk.

I think I knew it as a kid, but I'd get my earlobe stretched for thinking it, or heaven forbid, saying it. (Which I did to my mom...only once.) I thought he was a jerk 'cause he just couldn't see things my way; that of a teenager with a brain, an attitude and a dose of the '60's. He was so bloody conservative and unbending, it was either his way or no way.

Which leads me to the observation that the security industry is rapidly becoming a testosterone driven industry second only to professional wresting. Experts diss experts for having different opinions. Newly recognized twenty-something hacker/security people are thoroughly convinced their excrement has no odor. Vendors battle vendors as if they were defending their turf from imminent attack by the Death Star. Sorry, but I have a problem with that kind of attitude. Like my uncle, too many self-proclaimed security experts think that they have the only possible answer to security problems... and I laugh.

Security is more complex than any one organization, any one business process, or any one person's view or agenda. Security, as we have learned, is exceedingly complex. Recently I have been hearing from vendors new to the security field, "oh, our new blah blah product solves every security problem the planet has every known, and it's just $12.71 per copy. Would you like to write it up in one of your columns?" Security-dude Bill Murray once told me, "I know everything there is to know about computer security." After fifteen years in this field, I adhere to the following tenets:

1. No one product will today, tomorrow or ever, solve all of your security needs. Ever. Period. If someone tells you they have the panacea product, run as fast as your hoofers will carry you. It takes a highly coordinated suite of products, processes and people to construct a secure environment.

2. No one person knows it all. They can't. No way. If someone tells you they do, join the exodus from #1 above.

It takes a lot of folks, each with their own areas of expertise, to work together, not at ego-driven cross-purposes, to build a working security team.

3. Time Based Security is wonderful. Yes it is. But it does not replace everything we have learned. It is a new tool to be integrated into the security soup we batch up every day.

The emphasis that Time Based Security places on Detection/Reaction processes has pretty much concentrated on those detection methods that have already reached some degree of popularity, such as IP sniffing, perimeter penetration detection and host based audit analysis. And that is a good beginning.

Network protection, though, is about a lot more than just seeing if an acne-scarred, Jolt-drinking teenage punk is Push-Button hacking your front doors. It's a lot more than stopping an ex-KGB intelligence officer from stealing files. We cannot forget two other factors which are so often overlooked when building network security programs:

- **The insider**. The ex-employee. The disgruntled worker. Think about this for one terrible moment. Which of your staff do you provide the greatest physical access to your most sensitive areas, especially during low population off-hours? Who are the two lowest paid groups of people in your organization? The answers to both questions are the same: Cleaning staff and guards. It is so much easier to launch attacks when one has both trust and access; technical acuity not necessary.

- **The network itself**. Murphy's Law prevails at every turn. Equipment failures that affect network performance. Induced errors, internally or externally. Software glitches. Crash...sis-boom-bah!

System/Network Diagnostics

Large networks utilize an assortment of network monitoring tools already; it's part of the administrative job. When a network gets overloaded, from either extra-heavy traffic or a malfunction, which causes bottlenecks, a network monitor detects that event.

The administrator can then respond in whatever manner is appropriate.

However, these types of events are generally viewed as network management functions, not security problems. Time Based Security invites network performance and diagnostic monitors to complement other detection methods in gathering a more complete picture of the network. Then, when the monitor detects some abnormal behavior, the detection mechanism should drive the reaction channel, so that security can assist in determining whether the event was an operational anomaly or an attack.

Monitoring tools are effective at identifying software at nodes in the network and are often used for copyright/license compliance. However, the same mechanisms are applicable for identification of miscreant software at the user's workstation. Likely, you don't want users to have IP sniffing software, steganographic tools, their own encryption software tools or any of a range of hacker tools. By sweeping across your internal networks periodically you can identify those software components that violate your corporate policy.

Event/User Monitoring

Detection devices can be added at more nodes in a network to improve security, if there is an adequate reaction policy and mechanism in place. Monitoring decentralized nodal system activity can provide massive amounts of information to establish norms, trends, and systemic errors when the sampling is sufficient. It can also tell an administrator how often a floppy disk is used, for both reading and writing data.[47]

However, all of the data in the world is useless unless it can be analyzed and acted upon. For those of you who recall the PC-security business of the mid-1980's, these sorts of products were minor-league popular, but never exploded into widespread use due

---

[47] *In my humble opinion, floppy disks represent a huge security threat. Productivity falls when staff performs personal work on company time, brings data in and out or plays games. Viruses and malicious code are easily brought into a system, no matter now innocently. A floppy disk is also an easy and casual way to secret out private corporate data. Fair warning. I always recommend that companies seriously examine this overlooked vulnerability, and either remove them altogether or put a lock on the floppy drive.*

to complexity, incompatibilities and installation, configuration, and management costs. These types of products were viewed as local access control systems, not as a piece of a holistic enterprise system, much less a Time Based Security approach.

Enterprise Audit Trails and TBS

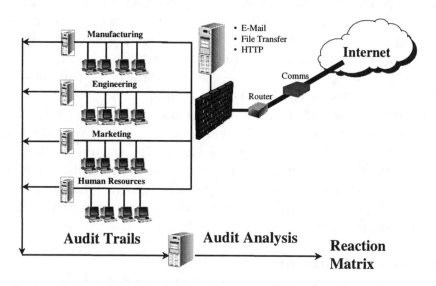

Keystroke Monitoring

If you can detect, store and analyze every single keystroke made at every single terminal in an enterprise, a tremendous amount of information can be gathered, not only with a security view, but also from a productivity standpoint. With the ability to track keystrokes at nodal terminals, individual efficient and work output can be measured, assessed and compared to other workers in similar positions. However, be very careful with implementing such mechanisms without the advice of human relations and legal counsel. The U.S. Justice Department has legal views that need to be considered, and states have their own interpretations which do not always favor a company's belief that it can control or monitor its own networks any way it sees fit.[48]

Collecting profiles and signatures of behavior and keystroke sequences creates a foundation for a detection mechanism. Just as some perimeter intrusion and host based detection systems (not to forget virus detection!) are library based; such is the approach to keystroke and event monitoring.

Contents Analysis

One of the more formidable tasks that the security industry faces is how to deal effectively with contents analysis. In short, the goal is to intercept data transmissions prior to arriving at their destination (or gateway), analyze the information contents for contraband data, and stop it. The problems to be surmounted are immense: how can vast amounts of traffic be halted, detoured to an analysis server of some sort, and then acted upon without causing severe traffic delays? Architecturally, where does a company wish to place such a contents analysis mechanism, and which information traffic flows are considered worthy of such an effort?

We are at the infancy of such technology, with contents algorithms having marginal successes in English much less other languages.[49] Greater CPU horsepower is also required to effect this approach without undue overhead. At a simpler level, the detection system could recognize encrypted messages and policy-choose not to permit their transmission. At the http server, a detection system, similar to Net-Nanny-like products, could look for objectionable content based upon key words. From the TBS standpoint, when such technology becomes more readily available and cost-effective, the detection mechanisms beg for a reaction channel output to thwart whatever activities are against policy.

---

[48] *Invasion of privacy is the counter-argument. Should an employee be given carte blanche to use a company's network without monitoring his actions? Why doesn't a company have the right to closely monitor its staff's keystrokes? Your company needs to meet keystroke monitoring head-on if it's a real consideration.*

[49] *Ways to get around any contents analysis include steganography. For example, if I modulate a single bit in a video signal, a compatriot could decode the message encoded into that bit. Remember, nothing is perfect. When such capabilities become commonplace, an organization could choose to ban all video material until solutions are found.*

Traffic Analysis (Behavior)

We all remember the Gulf War. My son, Adam, was born during the opening shots which my wife and I (mostly I) watched in the birthing room. Perhaps that is why we call him, and he acts like the Adam Bomb. (Or, it's 'cause he's real smart and seven years old!) Regardless, how did the media know that that war/conflict was going to start? After all, we had been building up our military capability there for months. What was the clue?

Pizza.

On the first night of the bombing campaign, pizza deliveries to the Pentagon increased something like 12,000%. Soldiers had to manage the war, soldiers have to eat and Pizza delivers. With Pizza Hut and Domino's trucks behaving like bumper cars in the Pentagon parking lot, the media got a pretty darn good clue something was afoot in Iraq. This is called traffic analysis.

On the network level, it's about the same. Say a network usually operates its T-1 to the Internet at 30% utilization, with bursts to 85%. Then one night, it sits at 72% for hours on end. If it were my company, I would like to know what the heck was going on. Wouldn't you? If Bob and Alice never talk to each other within the company network, yet over a one-week period they suddenly exchange 48 emails, something has changed. If John's profile says he rarely uses the Internet, but he suddenly is sending large amounts of data to SpiesRUs.com.ch (ch = china), as a manager I would be quickly suspicious.

In all of these cases, the suspicion is raised by behavior detected through traffic analysis, not the actual contents of the communications. Traffic analysis tools make an ideal detection mechanism if the baseline profiles are reasonably set, and the reaction channel can be whatever management chooses it to be.

Certainly in mission critical or security acute organizations, seemingly Draconian measures might be appropriate, and we find ourselves right back at the Reaction Matrix. For example, if a low usage traffic path is suddenly clogged with bandwidth-filling activity, an automatic response of shutting down that connection is not unreasonable. When a human is notified of the automatic reaction and then intervenes, innocent behavior can be quickly restored. Inappropriate behavior, though, has been thwarted, and damage avoided. With internal networks, and strong comprehensive detection and monitoring systems in place, the perpetrators are also identified.

Steal This

Billions of dollars in information are stolen every year.

Safeware Insurance Co. of Columbus, Ohio., reported 411,000 computer thefts in 1996. Hardware losses totaled $1.5 billion and the value of lost data was estimated at $15 billion. Twice as many laptops were stolen as desktops, and the number of laptops stolen increased 28% from 1995 to 1996 alone. What's more, 60% of computer thefts are perpetrated by corporate insiders: employees, temporary workers or contractors.

How can Time Based Security help us?

New technologies and products address the hardware thefts with software devices installed into the computers that expect legitimate users to perform certain identification procedures before using the machines. If the computer is stolen (especially laptops) and the thief attempts to use the machine without the right codes, the computer detects this error and takes a reaction. The preferred reaction is to "phone home," a central monitoring station, which hears the cries for help from the stolen equipment and then invokes a physical response. These new emerging business models anticipated Time Based Security and the shift to rapid detection/reaction mechanisms as a replacement for exclusive Fortress Mentality.

Detection on Protection Mechanisms

As has been suggested in the last couple of chaplets, protection mechanisms can be enhanced to provide better defenses. When protection products integrate detection/reaction channels, the overall state of network defense will rise significantly.

No one security approach will solve all of your security problems and not every security approach is right for everyone. The judicious choosing of products, processes and people is the only thing that will ultimately serve to provide secure environments for networks, enterprises and infrastructures.

Next, we want to briefly examine what more the vendors can do to enhance their product offerings, and assist us all in our respective security efforts.

25
Specifying A TBS Metric

Liza Minnelli and Joel Gray proclaimed that *"Money makes the world go around..."* in *Cabaret*, which half a century later is even more true in a global economy. But today, vendors, products manufacturers, the software industry also makes the world go around.[50] Software defines our views of the world, and in the case of security, what products the manufacturers choose to introduce drive markets and buying sprees by terrified network managers. And so it is with TBS; a few forward thinking security vendors are going to continue their market leaderships roles and begin building Time Based Security components into their philosophies and products. It won't happen overnight, but it will happen because the market will demand it.

For Time Based Security to proliferate, an adequate benchmark reference system is required. I have so far avoided the temptation to do more than explain the process and philosophy of the proposed model. But the astute reader will have already seen a problem with the assumptions made here: I have not qualified or quantified **how** we measure time. We live in a heterogeneous world, therefore, interoperable reference points, or baselines are required. Two possible routes come to mind, but other approaches may vie for position as the vendors take to benchmarking their own products.

1. If we stick with a 'real' time system, such as with microseconds, seconds and minutes, systems measurements would need to be on a common frame of reference. Thus, benchmark performance must relate to a specific hardware platform, (i.e., Pentium II 500, NT 5.03b, 128MB/RAM, etc.), as it so often does with other applications. In addition, communications and network bandwidths must be specified for many of the enterprise-wide views. Comparison and translation of system specifications on so many different heterogeneous

---

[50] *Just ask Congress and the Justice Department after their 1998 encounters with Gates.*

platforms and environments requires benchmarking and a means of relative translation between the different platforms.

2. The second possible approach is to make measurements in clock cycles. This is a good solution for comparing different manufacturer's software solutions on a single platform, but a means of heterogeneous integration of time measurements will still have to be performed.

Both methods have advantages and weaknesses.

The advantage of (1) is that we all seemingly understand time, but I would defy anyone to experience an emotional reaction to the difference between 1 microsecond and one millisecond, even though they are three orders of magnitude off. We'll see ads like,

"I can detect that ping in 491 nanoseconds. You took 1.45 microseconds."

Leave that to the technicians.

However, on a macro enterprise scale, the CFO or CEO of a company can much more easily relate to the following: *"You have a choice: protect your information systems so they are only vulnerable for either a period of three seconds, or for ten minutes. Here are the measurable potential losses for each instance. It's your choice."* Or, *"using the TBS model, in our existing network, we know that the following critical files are potentially vulnerable. Here are our measured options...."*

The auditor should have performed the analysis by examining the **E** time-based exposures and relate them to the amount of damage that can occur and the estimated cost of that damage.

This will have more meaning to the non-technical manager than saying, "the protection offered is either 120,673,213 clock cycles on a Pentium II – 300 or 233.4 million clock cycles on an Alpha. The choice is yours." Most of us cannot intuitively 'grok' the clock cycle rate on either machine, and translate.

So, this naturally brings up the issue of specifications, measurement, competing manufacturers and a lack of standards. Security manufacturers are going to be asked by their potential customers to specify performance. The makers of detection and reaction products are going to be leading the pack with real-world speeds as a measure of competitive advantage. Protective products will soon begin building in the Time Based Security components to increase their efficiency and effectiveness. The security manufacturers are going to need to measure their system performance in a number of ways.

1. Define the attack to be detected. An even playing field for the manufacturers and honesty for the customers. We have an excellent model for this; the virus industry and standard measurement techniques. In the coming months, we hope to begin publishing such specifications. Thus, attack detection/reaction standards for a number of detection suites for the industry are a viable and attractive short-term goal.

2. The amount of time it takes to detect attacks, from the moment the first attack code reaches the target to the time it provides an output 'flag' to commence a reaction.

3. A confidence level. False positives and false negatives are important to consider, especially in gray decisions and heuristic systems. Systems should be tunable for the administrator to choose on which side of the error-line he wants to function. We are back to policy here, but the percentage confidence level will need to become measurable criteria for these sorts of products.

4. The amount of time it takes to create a specific, well defined reaction to that specific detected event. The reactions are not defined by TBS, but the reaction chosen for any specific event must be quantitatively measurable and repeatable. It might be that a detected attack has several reaction options depending upon other criteria. Each (**D + R**) relationship pair must be specifically defined as part of the Reaction Matrix and to measure overall security strength.

5. Constantly update their products with complete TBS specifications as new attacks are discovered.

No matter that there's a driving need for benchmark performance criteria, we should also be somewhat suspicious of 'specsmanship' on the part of security products vendors. Therefore it will incumbent upon the security practitioner to perform system analysis and measurement of the integrated protective security shell and determine whether the products perform as expected. For the first time, then, the security industry will:

1. Have a quantifiable method to specify and measure the performance of security products by vendors and integrated security processes in the real-world.
2. Design by a quantifiable method, a predictable efficacy of protective security.
3. Assess hard dollar exposure to both protected systems, unprotected systems and TBS designed systems using conventional risk analysis tools and techniques.
4. Permit the client to verify the security performance.
5. Determine potential vulnerabilities via the TBS formula.
6. Limit risk through enhanced multiple protection layers.

The following chart is a rough example of how an administrator wishing to add protection to his network assets can urge manufacturers to comply with TBS measurement criteria in a benchmark environment. Magazine and independent lab testing will also find that TBS will assist their readers in picking equipment. Remember, though, that once you have a sub-system deployed in your real networks, run tests to verify quantify the TBS performance.[51]

Please recognize that these charts and mechanisms are our current thoughts and ideas, nothing that is mounted in concrete. They will hopefully serve as a starting point for vendors to quantify the security functionality of their products, and for customers to

---

[51] *I used to work at the defunct KayproComputers in the mid-80's. They never tested a completely assembled machine, thereby having the highest failure rate in the civilized world. The Kays' philosophy was to test all incoming parts, and if they worked, so must the complete computer. That reasoning ended them up in bankruptcy. Now that you have been offered the means to quantify security, do it. Test it before you begin, after its installed and periodically as part of ongoing security integrity checks.*

Product Comparison	Date:	___/___/00						
			Product #1		Product #2		Product #3	
(Detections Only)	Desired	Conf	Actual	Conf	Actual	Conf	Actual	Conf
Detected Event (Anomaly)	Time	Level	Time	Level	Time	Level	Time	Level
3 Bad Password Attempts	1 sec	100.0%	2.4 secs		1 sec		0.4	99.5%
3 Bad Password Attempts	1 sec	100.0%	.94 secs		1 sec		0.4	99.5%
Mulitple Port Scan	250ms	100.0%	1.5 secs		250ms			
Internal User - Audit Bahavior #1	10 secs	95.0%	95 secs	98.0%	110 secs	99.0%	5 secs	93.0%
Ping of Death	500ms	99.0%						
Syn-Ack Attack	5 secs	95.0%	2 secs				10 secs	99.0%
Mail Bombs		99.0%						
Firewall Breach Attempt	100ms	90.0%	2.7 secs		100ms			
Traffic 2X Anticipated		100.0%					3 secs	100.0%
Multiple Site Attack	3 secs	90.0%	2 Days		3 secs			
Shut Down $ Server	1 min	99.0%	2.4 hrs		1 min			

use as an additional gauge of what they are getting for their money. (If you have suggestions for the vendors, please let us know, and we will get them published on *Infowar.Com*.)

26
Surviving
Denial of Service[52]

Overview

"The Internet infrastructure lacks basic mechanisms that have been present and successfully used in telephone networks for a long time."[53]

As business models and global high value commerce migrate to the Internet, security concerns become paramount. While cryptographic techniques are able to address confidentiality, integrity, identification and repudiation issues, availability has been largely ignored outside of classic disaster recovery applications. On the Internet, denial of service has proven to be and will further become a major problem for both existing and evolving services. This paper suggests a model by which certain Internet based Denial of Services attacks and events may be successfully thwarted. [54]

Confidentiality and Integrity: Solved

The classic information security model in use today is generally based upon a triad of concepts:

- Confidentiality, or keeping secrets a secret.
- Integrity, or insuring that data is not altered without detection

---

[52] *The original version of this chapter was a paper presented at the International Baking and Information Security (IBIS) Symposium, February 1997, and the NIST/NCSC Computer Security Conference, October, 1997..*
[53] *"Analysis of a Denial of Service Attack on TCP," Christoph L. Schuba, et al., COAST Laboratory, Perdue University, May 1997*
[54] *I would like to thank Dr. Gene Shultz of SRI in Menlo Park, California, Dr. Gene Spafford, Perdue University, Robert Ayers from the Defense Information Systems Agency and Jack Holleran of the National Computer Security Center for their encouragement, support and comments.*

- Availability, or making sure that systems are "up and running" when they are needed.

The later, and under-recognized work by Donn Parker of SRI, attempts to expand this concept into a hexad of fundamental security principles:[55]
- Confidentiality + Possession
- Integrity + Authenticity
- Availability + Utility

By using strong cryptographic tools and well managed key-management structure, confidentiality of varying degrees of efficacy can be achieved to prevent message contents disclosure. A Sunday paper cryptoquote might keep a seven year old from reading the contents of a message encoded with alpha substitution and 40 – 56 bit crypto is subject to a successful brute force attacks. To gauge relative cryptographic strength, increased key length and implementation sophistication must be measured against the resources of ones adversary.

And so it goes for integrity where the premise of the solution is within cryptography. While there is substantial room for improvement in ease of use, scalability and wide spread implementation of integrity functions, the basic technologies are available, well established, reputable and can serve in key management, user authentication and non-repudiation applications.

Availability and Denial of Service

In March of 1995, I wrote a non-technical article, which explored Denial of Service attacks on the Internet as a vehicle for voicing political, social or anti-social discontent.[56] It claimed that Denial of Service would become the hostile remote attack weapon of choice in the near future. An expanded version of that article was

---

[55] *"Demonstrating the Elements of Information Security with Threats,"* Donn B. Parker, SRI *Int'l. June 1994.*
[56] *"Cyber-civil Disobedience,"* by Winn Schwartau, *Information Week, March, 1995.*

subsequently published,[57] and further examined the use of Denial of Service (DoS) attacks as a 'weapon-system' for which counter measures must be rapidly developed.

As predicted, in November of that same year, a so-called Internet strike was waged by dissatisfied citizens of France against government Web sites, shutting down selected services for the better part of a day. A similar assault was waged in Italy, the following month. The first large-scale media-grabbing DoS attack in the U.S. struck Panix, a New York based ISP in September of 1996. In March, 1996, an attempt to hamper Mexican government services was effective at shutting down some banking operations.[58]

Denial of Service indeed, represents the easiest, most effective way to cripple organizations whose connectivity is crucial to their operation and survival. The distinguished researcher William Cheswick agrees that DoS is a significant challenge facing the security community. "This [the Panix Attack] is the first major attack of a kind that I believe to be the final Internet security problem. We're going to see a lot more of this." [59] Years ago, Steve Bellovin published a highly detailed technical description of various DoS attacks, similar to those we have been witnessing in the last year or so.[60] Recent discussions on the Net and within the business community have finally recognized DoS as a serious threat worthy of discussion – not one to be ignored.

Denial of Service attacks can take many forms.

At the infrastructural level, The US Air Traffic Control system has been struck hard with DoS events, albeit not from Internet related circumstances. Due to the old (some say antique) equipment in use, lack of redundancy and reliance upon external communications and power systems, DoS events have been common. In 1995, dozens of air traffic centers failed for a combined

---

[57]. *"Information Warfare: 2nd. Edition. . ." by Winn Schwartau, Thunder's Mouth Press. 1996 ISBN 1-56025-132-8*

[58] *Denial of Service events have been plaguing all sectors of society. Airline DoS events consistently grab headlines, as do power outages. The increase in DoS events is alarming, yet satisfactory responses are as elusive as ever.*

[59] *Wall Street Journal, September 12, 1996.*

[60] *"Security Problems in the TCP/IP Protocol Suite," Steve M. Bellovin, the Computer Communications Review, Vol. 19, No. 2, pp.32-48. April 1989.*

thousands of hours, occasionally stranding airplanes without ground communication.

National headlines describe massive power outages and telecommunications failures that have infrastructure rippling effects, affecting tens of millions of people instantly. AOL and other portions of the Internet have also experienced wide-spread outages; DoS events, not attacks. The security industry is waking up to the unpleasant fact that significant portions of our critical infrastructures, along with the Internet, are vulnerable to DoS, which must somehow be balanced with protective or recovery methods

Over the years, the disaster recovery business has flourished. It has proven itself to function very well in the face of electrical outages, hurricanes, floods and other so-called Acts of God. We are acutely sensitive to distant events as Miami's landscape is flattened yet a data center located there is subject to only minor hiccoughs. And even though Chicago and New York flooded, our national economy didn't suffer. These are the same arguments that my occasional debate partner Martin Libicki[61] when he suggests that Information Warfare techniques (cyber-terrorism, civilian infrastructure damage or collapse) are not likely to cripple to the United States any time in the near future. After all, he suggests, we've been through it before and we've survived.

That is partially true. We have well survived local events. Like Hurricane Andrew. Or the San Francisco earthquake (or LA for that matter). We survive them well, because they were local events, and resiliency is the product of built-in redundancy of the localized transport layer. We survive them because they affect relatively small numbers of people, for limited amounts of time.

As significant portions of the US and global economy rely upon distributed Internet based communications for business, revenues and profits, the concern over DoS and solutions implementation rises dramatically and profoundly. Remote banking and international commerce knows no geographic bounds, and thus

---

[61] *Formerly of the National Defense University, Washington, DC.*

we leave the realm of the mere local and can see more widespread effects.

Financial institutions are increasingly looking for means to decrease operational costs, increase profits and bring additional remote services to home and business. [62] According to the ABA, 50% of current retail banking presence will go the way of the stagecoach and be replaced by electronic banking. Through Web sites and email, a customer will either direct link over the Internet or in some cases, special client software might be used. With the availability of quite acceptable security techniques, the issues of confidentiality and integrity are essentially solved. The Airline industry is moving to Internet based electronic ticketing which reduces their ticket processing costs over 90%; from an average of $8 per ticket to $.50 each. Any interruption to this service directly affects bottom line profitability.

However, because the historical thinking of the professional security field has not adequately dealt with the issues of DoS as a function of the Acts of Man, we will, as William Cheswick also suggests, find an increased number of intentionally triggered hostile DoS events with much more far reaching effects than we already have seen.

It is the Acts of Man, intentional denial of service attacks, which have so far been very successful and represent new challenges to the Infosec community. Thus far, little progress has been in solving this daunting problem. But, we hope to shed some light on the approaches necessary to deal with this third and final triad of info-security.

The Simplest Place to Start

Let's begin our search for a solution by looking at connectivity as is typical today, but in an idealized scenario with only two and three computers respectively. The most prevalent connectivity models in the current and planned future use employ a

---

[62] *In private discussions with global financial organizations world-wide, Denial of Service produces greater fear than the occasional hacker or embezzler for which they are reasonably well prepared. The issue of public confidence in the institution itself, as well as the banking system, becomes more important than comparatively minor losses from transaction errors. Denial of Service is the electronic equivalent of shutting a bank's doors in their minds.*

single 'wire' along which both information, transport and control signals travel. This 'wire' is both logical and physical and TCP/IP represents one such realized implementation.

The logical 'wire' is how most Internet traffic and commerce is handled today and for the foreseeable future. Which specific physical wire is used to get the information to a specific destination is immaterial in the TCP/IP model, as long as it gets there. Also, the control signals must communicate properly between the right pieces of hardware and software, and it, too, is blind to which physical 'wire' it uses to accomplish it's task.

And therein is the crux of the problem – and the tantalizing attraction of TCP/IP and similar protocols. As below, two computers connect on a single physical and logical wire. If that wire is cut at any place between the two computers, all information and control communications are severed as well. However, in an Internet-like connection, if the physical wire between any two machines is cut, communication can still occur between all three members of that network.

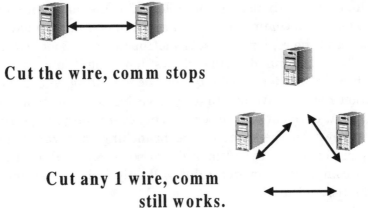

Cut the wire, comm stops

**Cut any 1 wire, comm
still works.**

With the 'Net' or any generalized Public Service Network, again the niceties of IP come into play. If a portion of the Network Cloud is out of service, communications can generally take place, but in reality, there may be something of a performance hit, such as a time delay due to increased density of traffic at certain router locations.

Let's Complicate It A Bit . . .

However, communications to end points occur in the so-called 'Last Mile,' such as the dedicated piece of copper wire that goes to individual homes, or thin fiber links that connect larger businesses to the 'Cloud', represent the physical weak spots. If the Last Mile finds itself severed, the 'Cloud' will function, but communications to and from the severed link will not take place.

Ron Eward's seminal work into transport layer vulnerabilities underscores the weaknesses of a global economy dependant upon the consistent and reliable operation of a few dozen strands of glass cable, less than a millimeter in diameter. While I have used the term 'mappers' to identify those people who have created a Rand McNally road-map of Cyberspace,[63] the fundamental premise of Mr. Eward's work is that connectivity requires real world, tangible hardware and that 'Cyberspace' is not really as ethereal as is generally perceived. High speed, high bandwidth routers are connected by copper and fiber optic wire. The hundreds of thousands of miles of cable and wire are the backbones of the communications grid; millions more make up the sub nets, and the regional nets, or metropolitan area networks.[64]

The following drawing shows how the Cloud has spurs, which branch out for flexibility, physical routing and bandwidth utilization reasons. We see that by severing a connection from the Cloud, we at once cut off the services of several large organizations. Further down to the right of the branching, a severed line will affect pairs of firms, and ultimately, at some physical location near a target company, a severed cable will only isolate that one target from the rest of the 'Net.'

---

[63] *"Information Warfare: Chaos on the Electronic Superhighway," Thunders Mouth Press, 1994*

[64] *"Telespace, Telewar and the Vulnerability of a Global Electronic Economy," by Ron Eward, March 11, 1995, Martech Strategies. Presented at InfoWarCon V, September 1996, Washington DC and InfoWarCon VI, May 1997, Brussels, Belgium.*

DENIAL OF SERVICE AT THE PHYSICAL TRANSPORT LAYER

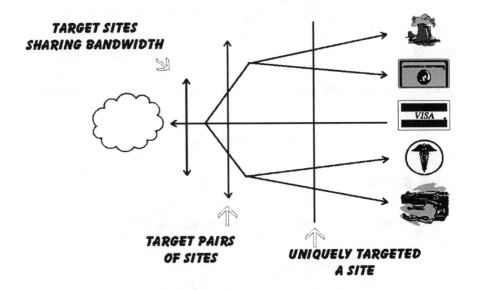

The lack of physical redundancy at the transport level echoes a similar TCP/IP deficiency exacerbated by remote software based DoS attacks: lack of logical control signal redundancy. While cutting the physical wire to a target host or organization is certainly effective as an attack technique, there is are significant drawbacks due to the physical nature of the attack:

- one must be physically near a specific target, or
- one must physically expose oneself to launch a physical attack, thus increasing the chances of being caught.

Remote, non-physical attacks represent a step forward in sophistication and anonymity. Thus, electronic mail bombing is a good alternative for the so-called Information Warrior.

If My Hose Is Bigger Than Your Hose, You're Hosed

The nature of email bombing is such that anyone can attack anyone else, and it's all a matter of size. Most folks have a 28.8K (or similar) modem at home. Businesses have anything from an ISDN to a T-1 or T-3 which offer tremendous bandwidth for large numbers of users.

If an ISDN source, with a constant data stream is sent to the target 28.8K modem, overload is the result. Just think when a user connects to his personal ISP, and he is receiving a huge file; all other traffic is blocked until that transmission is complete. So, a single ISDN can send continuous traffic to a single 28.8 connection and dominate it.

D.O.S. HOSING

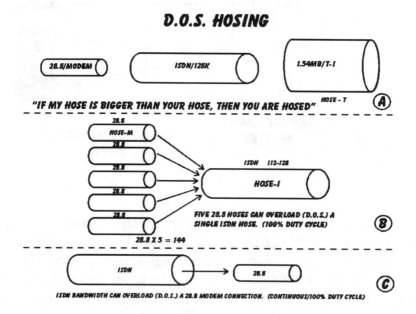

Even with the splintering of files in IP, under 100% duty cycle transmission, a series of a smaller hoses, can overload a bigger hose (5X28.8>1-ISDN) thereby cutting off the larger hose's ability to communicate. Please note, however, that hose size is not important for many other types of DoS attacks, such as Syn-Ack.

The following shows a freely accessible on-line E-mail Bomb service. [65] Merely choose the target address, enter in the 'Terrorist Demands' and choose how many thousands or millions of email

---

[65] The original location of this site was outside of the United States. With the help of the national police of that country, the site was closed, but not before being extensively mirrored.

bombs to send to the victim. The beauty of this technique from the attackers' viewpoint is that through built-in source address spoofing, anonymity is maintained.

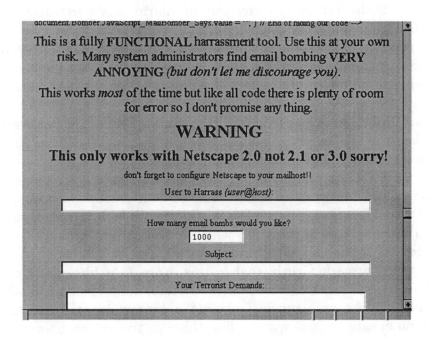

Denial of Service attacks take many forms. As more and more tools have been developed, DoS strikes increase. Infinitely Pinging a target from a number of confederates is a moderately explored method, as is sending endless streams of giga-byte length messages.

A targeted email bombing DoS attack, effectively isolates the target from the rest of the Net for as long as the attack lasts. However, depending upon the nature of the attack and the relative bandwidth utilization, the ISP may not be aware that one of its downstream customers is even under assault. The ISP may not notice that certain unfriendly activities are taking place, unless, its own services are hindered or overloaded – or, as we will show, even has the necessary Time Based Security processes and procedures in place to detect malicious behavior of any sort. On the other hand, if the attack represents a relatively large percentage of the local ISP's traffic, not only will the victim be affected by a successful bandwidth-filling email bomb attack, but the performance of the ISP will suffer, too. Thus, the service to its other customers will degrade through collateral damage.

If the target has any sort of Alarm or detection system, which notices that it is under attack it must have the means to respond to the attack. Assuming that the target has a response system as well, many DoS attacks will simply overload the victim's ability to communicate back to the first-hop ISP or even internally to its own servers. Thus, if its internal security system attempts to alert its administrator by email, that call for help might be severely delayed or the message might never get through at all due to the bandwidth filling or control signal confusion.

So, we should no longer view the attempted transmission of the Call For Help as an email message with content. Functionally, for modeling sake, we can view the Call for Help as a control signal – a warning light if you will - which in the current protocol suites, attempts to use the same paths, routing and physical connectivity as the information path (that in this example is blocked). Herein the attractiveness of the TCP/IP protocol fails. Similarly, when other attacks are launched against a target site, the means to communicate back down the line through the Internet, even one hop, may be thwarted by bandwidth filling, port overload or other disabling effects.

Tracing back through the multiple hops of the Internet provides several advantages, including identifying the specific source of hacking. Certain firewall manufacturers have claimed to be able to 'strike back' at an offending intruder or attacker, but without widespread protocol cooperation, such efforts are extremely limited.[66]

Regardless of the nature of the attack, from the victim's viewpoint, the results are the same. A server or client is overloaded and/or isolated by a particular technique

---

[66] *Secure Computing Corporation, Sidewinder Product brochures, October, 1995.*

Adding a Protection Module

.In the following diagram, the top network is being targeted by email bombs from a single source. Now, let's add Detection. This is an 'intelligent' device which serves as an off-line real-time detection-monitoring-alarm system. In this case, the Detection system employs IP-sniffing as well as system behavior monitoring from an audit component located at the host. (Detection should logically sit on the side of process and does not interfere with information flow. [67]

Detection receives audit information and real-time information from the target's node, servers, and its sniffer component, and than examines the database upon a wide selection of criteria to be chosen by the administrator. A sampling of some criteria for analysis might include:

- Source address (versus time); how many emails from a single source address in a given time period.
- Source address (versus content length); normal range of size of emails per time period.
- Syn request (versus time before Ack/No Ack)
- Syn request (versus source address) per time
- E-Mail quantity (versus history/time) per time

The administrators initialize the parameters of the rules based engine within the Detection Mechanism by choosing those activities he knows he wishes to block at all times, similar to what is done with a conventional firewall. The nature of the 'engine' can range from static programmed parameters to dynamic heurism which 'learns' the system norms over time. Functionally, for this solution-model, the choice of that specific implementation is immaterial. The Reaction Matrix could include:

- Email administration of the problem
- Heightened security auditing at the target
- Telephone Call
- Alert a CERT
- Shut down connection to Internet

---

[67] *While logically it is software, in practice the Detection Module best works as a separate hardware device with connectivity.*

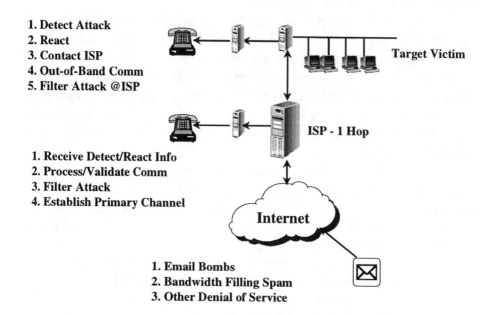

1. Detect Attack
2. React
3. Contact ISP
4. Out-of-Band Comm
5. Filter Attack @ISP

Target Victim

ISP - 1 Hop

1. Receive Detect/React Info
2. Process/Validate Comm
3. Filter Attack
4. Establish Primary Channel

Internet

1. Email Bombs
2. Bandwidth Filling Spam
3. Other Denial of Service

In the last option above, the remedy is not better than the disease – and in fact meets the original goals of the attacker anyway; service denial. These responses to DoS attacks represent conventional wisdom, but, none of these solve the Denial of Service while it is occurring. For example, Panix in New York knew very well it was under attack. It knew, as did its 6,000 customers, that they were effectively out of business until the attack stopped. Merely knowing one is under attack does not defend against it or stop it. Something else is needed.

The problem in the TCP/IP suite is that control and information signals co-reside on the same logical and physical topography. Thus, DoS attacks can often succeed no matter what detection mechanisms might be in place.

Control Versus Information

The original survivable goals of the Internet are different than they are today. "TCP/IP was designed to be resilient under failure, not under attack," said Dr. Gene Spafford. Thus, it made good sense for that experimental computer network nearly thirty years ago to combine control and information on a single path. However, when we look into other disciplines, we see that control and information signals are handled quite differently.

The transistor and FET (Field Effect Transistor) are the building blocks of our digital infrastructure and employ distinct physical paths for control and information flow. Voltage Controlled Amplifiers, common from DC to light applications, also separate control and information signals. So why not the Internet?

The problem is clearly one of installed infrastructure base; no model, no matter how good, will supplant the existing installed infrastructure – the financial cost is too great and the techno-political hurdles are immense. But, as we look further into the situation, we see that a full-time separate control signal and/or channel can alleviate much of the DoS problem. The need arises when the primary channel is too clogged or otherwise hampered from communicating.

Involving the ISP

No one person or domain on the Internet stands alone. The Internet is an integrated cooperative whole and requires agreements on interoperability, protocols and other evolving technical standards. From a social standpoint, acceptable behavior has developed on an ad hoc basis. This cooperative interaction has been the cultural and technical basis of the Internet since its 'survivability' genesis in 1969.

To date, the relationship between the user, corporate or individual, and their ISP has been fairly laissez-faire, similar to what many of us have with our telephone service companies. If there is a problem, we call them up, complain or speak with a technician, but otherwise, the relationship is, appropriately so, invisible. An ISP is similarly viewed as an invisible supplier of bandwidth as long as the communications are reliable. This is the historical model of communications, providing autonomy to the customer.

However, Denial of Service creates a different situation where the historical relationships may no longer be applicable as Dr. Spafford's opening quote suggests. DoS is a security concern unlike strict confidentiality or integrity measures, which are implementable on an end-to-end or point-to-point basis, independent of and transparent to the bandwidth provider.

To solve DoS concerns, a more tightly integrated relationship between customer and bandwidth provider or ISP is required. For a complete, defensive posture against DoS to be truly effective over the long term, we must involve most of the Net community globally. We will need to come to agreement on detection, reaction and interoperability to truly solve DoS on the current incarnation of the Internet.

The ideal goals to defend against Denial of Service attacks at either scale are fairly clear-cut, but lofty.

1. No modifications to existing protocols or altering the infrastructure.
2. All DoS attacks should be detected.
3. All DoS attacks should be thwarted.
4. False positives should approach 0%.
5. Recovery from DoS should be as rapid as possible.
6. The perpetrators of the DoS attack should be identified.
7. The defensive solution should generate forensic data for possible future prosecution.
8. The solution should be low cost.

A Model for DoS Defense

Let's add a corresponding component at the ISP, which we call a Reaction Module. Within the RM are Dynamic Filters which are closely entwined with the ISP's routers and switches' functionality. Based upon input from the ISP's own switches and from the associated Detection Mechanism's, the RM makes routing and filtering decisions which in turn, cause the ISP to either permit or deny traffic to pass through.

For example, if the Detection Mechanism (or Detection Module) analyzes audit behavior from the target network, "sniffs" its own IP traffic, and determines that an out of bounds condition exists, policy will probably dictate blocking the offending traffic. So, in this case, say email bombing is occurring from a single source, the source IP address can also be blocked at the ISP. Therefore the DM will attempt to communicate with the RM, which in turn tells the dynamic ISP located router filters what actions to take. But none of this works if the primary TCP/IP channel is crippled as is the nature of many DoS attacks.[68]

Referring to the prior diagram we add a control channel to permit communication between the target Detection and downstream Reaction in question. So, let's add a new path, called the B-Channel. The B-Channel is any communications path that does not use the same logical or physical connections as the primary information path as the DoS attack, such as the Internet. Using a telephone circuit is possible, however, it may suffer from physical attack as well; therefore, for generalized purposes, the use of cellular based modems is suggested for stringent requirements. They utilize an entirely separate infrastructure and contain no physical overlaps with the primary communications path.[69]

Operationally, the following sequence meets with a lot of approval from network and security administrators.

1. The Detection mechanism at the target detects the attack.

2. Assign a criticality value to the event detection (i.e., 1-5, where 5 is the most critical.)

3. The DM could choose to act like a personal firewall, communicate with the target host and dynamically filter out the offending information streams at that point, given such capability. Adding TBS mechanisms into Protection products is a good architectural (and sales) approach. For low density DoS attacks, this might be an effective method.

4. At the same time, the Detection Module at the target will attempt to communicate with the ISP's Reaction Module along the primary communications path. Automatic authentication and non-repudiation mechanisms will assure that messages sent by the DM

---

[68] *In this case, the $P > D + R$ (etc.) still provide the same value in considering the 'nakedness' of the network. Solving DoS with the cooperation of the ISP is a Reaction to the original detected event. This operational concept can daisy-chain throughout the Internet.*

[69] *The contents of the B-Channel conversation between the DM and RM is low bandwidth, and carries instructions, based upon the DM's analysis, requesting that the RM take certain actions.*

were in fact received by the RM, and that they are accurate and private.

5. The RM will use the information from the DM, in combination with the ISP's own software to make a real-time dynamic filtering decision to isolate the victim or block the offending DoS transmissions; whichever is the most effective reaction within the allotted time period.

6. If the DM does not receive an acknowledgement from RM along the primary communications path within an expected period of time,

7. Re-evaluate and upgrade the criticality value, and

8. The DM will initiate a B-Channel communications. The DM and RM converse, and then the real-time dynamic filters take over, ending the DoS attack.

Securing the B-Channel

The B-Channel offers a secondary target, especially for motivated adversaries, so it cannot stand unprotected as a defensive mechanism.[70] But, providing adequate security to the B-Channel is not terribly difficult. There is a low population of possible B-Channel paths to and from any one ISP or routing location; they are known and finite as are the TCP/IP routing paths of the Internet itself. The B-Channel is effectively a virtual topological mirror of the portion of the Internet we desire to protect from DoS. It is comparatively simple to secure a few known relationships.

1. All B-Channel paths to be opened only upon rapid remote encrypted, one-time authentication techniques. If the communications link is not established within a small window of time, the channel should immediately be shut down.

2. Upon opening the B-Channel with successful authentication, non-repudiation mechanisms insure that communications are complete and acknowledged.

3. Integrity checking for all communications.

4. Message contents, in both directions, encrypted.

---

[70] *Recall how defense in depth and protecting the reaction channel is a key TBS concept.*

5. Complete logging of all events and traffic along the B-Channel at both ends.

Getting to the Source

The primary goal in fast recovery from DoS attacks, of course, is to keep the target network functioning at near 100% reliability. And, though the B-Channel approach does effectively isolate a target from its attackers, determining who the attackers are has not thus far been considered a high priority. Commerce reconstitution is the first order of business.

As the Internet becomes victim to more and varied DoS attacks, merely stopping them will be insufficient to meet the real world desires and needs of the Net Business community. DoS grates at the reasonable man's sensibilities, and our sense of fair play and justice demands more. The contemporary criminal hacker gets away with electronic murder, so to say, because he can commit his particular sins with virtual anonymous impunity. "As long as a significant number of sites can transmit packets into the backbone without any source address checking, hosts are still subject to untraceable attacks." [71]

What we can do, though, with the model presented herein, is trace back to the source, hop by hop.[72] So, in the following diagram, the top network is under an anonymous DoS attack. Its Detection Mechanism detects it, and, in combination with the RM at ISP1, the attack is shunted aside, but the attacker does not know that he has been detected.

The original victim is functioning again, but now ISP1 with its RM send information back to its small but known group of routing partners one hop back and provide them with the 'signature' of the offending traffic or a last-hop ID.[73] Only one of the router partners will recognize and match the signature against ongoing traffic, and then it, in turn, will transmit the same signature one hop back again, closer to the source. As long as the

---

[71] *Schuba, ibid*

[72] *If the attacks are ongoing, this become a real time response. If, however, the attacks are finished, time consuming analysis of activity logs at the ISP will be necessary.*

[73] *It goes without saying, but to achieve this goal, the routing partners must utilize the same technical solution.*

routers in the attack path participate with a B-Channel model, the source tracing continues hopefully back to the source.

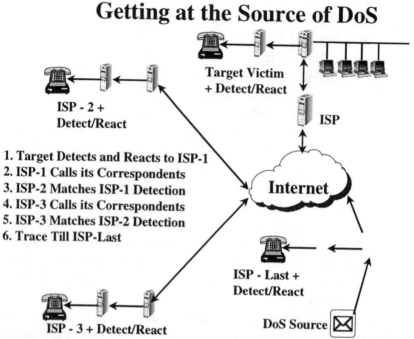

Getting at the Source of DoS

Target Victim
+ Detect/React

ISP - 2 +
Detect/React

ISP

1. Target Detects and Reacts to ISP-1
2. ISP-1 Calls its Correspondents
3. ISP-2 Matches ISP-1 Detection
4. ISP-3 Calls its Correspondents
5. ISP-3 Matches ISP-2 Detection
6. Trace Till ISP-Last

Internet

ISP - Last +
Detect/React

ISP - 3 + Detect/React

DoS Source

In Conclusion

For organizations which worry about DoS attacks, the proposed solution uses Time Based Security components and validations and it uses the networking principle for minimal partnering with other Internet entities. However, for complete Internet-wide DoS protection and attacker-identification, extensive cooperation is required on standards, protocols, security measures for B-Channel, criticality values and formatting. Another compelling feature is that this model offers complete co-existence with the existing Internet, not requiring any fundamental changes to the infrastructure of current standards.

Time will tell if the Internet and business community cares enough to implement solutions to this devastating vulnerability.

27
Infrastructure
Protection[74]

"As we finished the dive, streams of AK-47 bullets met us at the surface. I didn't call that peace. When the mortar rounds rained down around us, that was Peacekeeping. Not war."

US Navy Commander James K. Campbell,
Mogadishu, Somalia.

Preface (November, 1998)

Arguably, attacking a critical infrastructure is an aggressive act. Even if the attacked point of that infrastructure is a *mere network*, the function of the systems of the attack and our reliance upon their continued operation is what we care about. Let the electricity flow, pump the water, move the money, keep the phones talking and the planes flying. The networks have become the infrastructure itself: the information about them, their day to day operations and the controls for what ultimately is a huge hunk of steel propelling forward at 600mph in the upper atmosphere or a giant sluice gate falling into place.

This chaplet was originally written without TBS in mind, but as the concepts of both have evolved, it became quite apparent to many readers that there was a direct correlation. I intended to keep the body largely intact and use footnotes to make the TBS parallels, but I failed at that mission and did in fact amend the text throughout.

Since *Information Warfare* first appeared in 1994, I have attempted to place the protection of critical infrastructures as well as networks under the same umbrella and protection paradigm.

---

[74] *An earlier version of this chaplet was printed in "Cyberwar 2.0: Myths, Mysteries and Reality," Edited by Alan Campen and Douglas Dearth, AFCEA Press, ISBN 0-916159-27-2, available at www.infowar.com bookstore.*

The architectural comparisons are inescapably the same – just the implications are immensely more important to our nation and to other technically advanced countries with similar vulnerabilities. In this chaplet, I hope to show how Time Based Security can be used to increase defenses at both the enterprise as well as interenterprise/infrastructural level.

One of the great leaps forward we will all need to make, though, is to erase the isolationist in each of us, our companies, and the ever present interagency squabbling over turf rights. We will find that we first have to overcome our entrenched social stratifications, work together, and find common solutions that benefit all participants. After all, we ask all of our networks and infrastructures to work together. Why shouldn't we?

The Language of War

What exactly is war? and is our definition of war adequate to meet the challenges we face today? I spent my teenage years watching the Vietnam War, which in international legalese, was not a war. The Korean War meets the same linguistic manipulation, as does the Gulf War.

With all due respect to the tens of thousands of Americans who have made terrific sacrifices in the last 50 plus years, technically the United States has not been at war since World War II. We've had police actions, skirmishes, coalitions, and peacekeeping missions – indeed, we employ a plethora of Zieglerisms to define military intervention, military action and the death of American soldiers as anything but war. The word war has specific intent and meaning in international courts and to other nation states which we have diplomatically and politically avoided in the best interests of peace.

We sent the low flying US Air Force into Libya to bomb the bejeezus out of Khadhafi in 1986; we went into Granada and later tested the stealthy F-117's in Panama, but we haven't been at war. Thus, if the politicians fail, and if the diplomats fail, we send in the troops – but it still isn't war. Well, if it ain't war, then, what is it?

This conundrum is not an exclusively American phenomenon, either. The Soviet War in Afghanistan. The War in Chechnya. The Chinese troops march into Tibet. The Russian-Chinese border skirmishes. An examination of the history records from the last half-century will suggest peace, for lack of the formal declaration of War, yet millions of people have died during "Peace in Our Time."

Moving from Daniel Webster to a host of Thesauri, "War" offers quite a range of semantic alternatives:

- Clash
- Conflict
- Battle
- Contend
- Opposition
- Fight
- Attack
- Invade
- Struggle
- Hostilities

These terms are often bandied about as an alternative to the dreaded "War" word, but they also find themselves at home in marital dispute, corporate competition and children's games.

I find this a sad commentary, that while we have turned the killing machines, our soldiers and their weapons into high technology marvels, we have not similarly evolved our concepts of conflict and attempted to amalgamate the subtleties of hostility into our modern lexicon.

Into the Sublime

At the demise of the Cold War (there we go again), and as the US seemed to reign supreme as the sole military superpower, certain factions claimed the Peace Dividend as their own. Obviously, they said, the Pentagon no longer needed its inflated budget. One Japanese academic proclaimed the End of History. Others, though, saw the future not through rose colored glasses, but through transparent microscopically thin layers of silicon. A newer, friendlier and gentler kind of war was emerging.

- In 1991, this author appeared before Congress, and testified that "Government and commercial computer systems are so poorly protected that they can essentially be considered defenseless; An Electronic Pearl Harbor waiting to happen."[75] Information Warfare was no longer the exclusive child of its classified parents.
- Alvin and Heidi Toffler wrote about Anti-War as civilization moved into the Third Wave of an information-based society and how such capability would alter the face of war.
- Al Campen's *First Information War* examined the nature of high-tech conflict vis-à-vis the Gulf 'War'.
- The unclassified book, *Information Warfare* appeared in 1994 and for the first time described the nature of future conflict "without bombs, bullets or bayonets." The floodgates of offensive capabilities were no longer shackled by military or intelligence censors. Open source materials made themselves a thorn to some sectors.
- John Fialko's *War By Other Means* took the concepts of Class II Infowar established in *Information Warfare* and examined several cases where espionage was equated to war.
- Al Campen and Dough Dearth introduced *Cyberwar* in 1996, a compendium of thoughts on Infowar with a differing name.
- *Information Warfare: 2nd Edition* was published in 1997, and the combined works of fifty or so international authorities on Infowar contributed to this reference tome
- *Cyberwars: Espionage on the Internet* was first published in France and then the US in English as a specific form of Cyberwar – a definition still looking for acceptance.

---

[75] *I was immediately accused of "overstating the case," and much of the original research we did was dismissed, not to be confirmed and become mainstream for years. No one believed in cyber-terrorism back then.*

- The latest, *Cyber Mafias*, another French book further expands the Class II style of infowar from the organized crime viewpoint.

The generally agreed upon taxonomy for Infowar is:

> **Class I** – Assaults on personal privacy, psyops, perception management, the media and related area.
> **Class II** – Economic and Industrial Espionage
> **Class III** – Electronic Pearl Harbors, national assaults against critical infrastructures and rear echelon support systems.

Yet, despite all of these efforts, none of us have been truly successful in either defining or redefining war. So, is Information Warfare really war?

Alternative Semantics

The term Information Warfare is still anathema to much of corporate America and interested bystanders who understandably have trouble with the connotations of armed conflict. The term Cyberwar mollifies a few critics, but that nasty word 'War' is still nettlesome. The Pentagon, in defiant response to their own internal debates, is now floating the term Information Operations to explain what they do… but it still is far from as inclusive as many of us would like. The implication (in some cases as overt as tacit) is that the Pentagon is in the physical war business, and anything else is a peripheral adjunct to that prime directive. That contention, too, is a matter of healthy debate when we ask, "who protects the private sector from international assaults that do not involve bombs, airplanes and submarines?" To date, no one has successfully introduced a term that satisfies all people concerned.

If we look back fifty years to the birth of the US Air Force, when the Army as parent reluctantly allowed its offspring to fly on its own, we can gleam the depth of the complexity of adding a new dimension to warfare.

Regardless, in the hundreds of presentations I have given on Information Warfare over the last eight years, I get around this debate by asking my audience, "Can we all agree, that despite the fact we may disagree on what term is appropriate, we all do, in fact, know what we are talking about?" That consensus is generally a quick one – so we, my audiences and I - can get to the crux of the issues and belay the semantics arguments to a later moment. But now, let's do exactly that - address the semantics. Is Information Warfare really war?

I would argue that at certain intensities, where the goal of the actions taken is to completely decimate the adversary's infrastructures, be they civilian, military or rear-echelon support, we are about as close to war as we can get. But then, we have never been at war with Iraq: this is a case of semantics without resolve. In the ECO-D work I did for the military, the entire purpose of the offensive activities we developed was to get an adversary to act within certain limits, or to stop acting in specific ways. War.

Cyberwar? Is that war? Probably not a whole lot different than infowar above, but it may be perceived as less offensive.

Information Operations doesn't say much. InfoOps are support to the military in any operation, regardless of its intent or goal. Some analysts throw the offensive use of information systems into this category, but the semantic downplay of its offensive or defensive nature waters down any sense of resolve or national will. *"Don't tread on us or we will resort to Information Operations."* I don't think so.

Many of us have successfully made the argument that the economy is certainly a national security asset worthy of defense. But whose defense? Do we defend against 'cyber-attacks' upon US banks or the infrastructure? In these cases does a good defense mean we have a good offense as well? And what conditions need to be met prior to get involved in an Infowar or a Cyberwar?

That debate is not going to be solved here today, but the question underscores the need for an understanding of just how much resolve we actually have to defend ourselves and by what means and to what lengths we are prepared to go.

In presentations to military and political groups around the world, intelligence organizations, and to financial and corporate leaders, I have given them the opportunity to hypothetically decide how they should react in given scenarios. The consternation created ranges from amusing to disturbing because so few people have thought these problems through.

Now, please also keep in mind that our adversaries are quite different than who they were a few short years ago. The list is much longer, and the nature of the adversarial relationship is not nearly so clean-cut as 'War" and "Peace" in the Webster sense of the word. "The good old days" of the Cold War were so much simpler than having to codify and monitor the global myriad of adversaries.

- Private Domestic Economic Competition by US Rules
- Private International Economic Competition by US Rules
- Nation-State International Economic Competition by Other Than US Rules
- NGO International Economic Competition by US Rules
- NGO or Nation-State International Economic Competition by Other Than US Rules
- Nation-State International Economic Aggression
- NGO International Economic Aggression by Other Than US Rules
- International Economic Sanctions by US Rules
- International Economic Sanctions by Other Than US Rules
- Domestic Hackers – Domestic Military
- International Hackers – US Military
- Domestic Hackers – US Business
- International Hackers – US Business
- Domestic Hackers– Domestic Infrastructure
- International Hackers– Domestic Infrastructure
- Profit: Domestic Crime – Domestic Criminals
- Profit: Domestic Crime – International Criminals
- Profit: International Crime – Domestic Criminals
- Profit: International Crime – International Criminals
- Profit Oriented – US Criminals With Damage
- Profit Oriented – Int'l. Criminals With Damage
- Terrorism – Physical Destruction
- Terrorism – Psychological Damage
- Nation-State – Without Damage
- Nation-State – With Damage
- Denial of Service – Electronic With Damage
- Nation State – Attack on Military
- Nation State – Attack on Infrastructure

- Non-Nation State - (NGS) Attack on Military
- Non-Nation State – (NGS) Attack on Infrastructure
- Domestic Corporate – Domestic Military
- International Corporate – Domestic Military
- Domestic Corporate – Domestic Business
- International Corporate – Domestic Business
- Domestic Corporate – Domestic Infrastructure
- International Corporate – Domestic Infrastructure

Governments will certainly want to granulize this sample listing with greater, threat specific names. Concerned private organizations will probably also have their own lists of adversaries, ranging from domestic competition all the way to allied Nation-States targeting them for national security secrets.

The Options Chart

The range of available options and scenarios to given vulnerabilities and threats is best represented by a smooth continuum, with an infinity of subtleties. In practice, though, we have to quantize the threat. Chart #1 below is an example of how to define degrees of interaction between organizations we face today and are likely to face in the coming years.

On the left, create a range of possible interactions, from the best case (or no case) to the worst case, as suggested above. Then, across the top, choose a representative set of the possible quantifiable effects of that interaction. This is not adversary-specific, but by taking the your lists of potential competitors or adversaries, an options chart as below can then be created for each.

Losses	Corp $1M	Corp $1B	Corp $10B	Natl: $1M	Natl: $1B	Natl: $10B	Natl: $100B	Natl: $1T	1 Dead	10 Dead	1,000 Dead
No Contact											
Awareness											
Minimal Contact											
Simple Commerce											
Friendly Competition											
Defensive Allies											
Competition											
Overt Threats											
Economic Espionage											
Military Espionge											
Hostile Economics											
Physical Skirmish											
Economic Attack											
Military Attack											
Conventional Conflict											
Declared War											
Nuclear War											

Intensity/Goal

Now filling in these charts can be done in many ways, and the "On a scale of one to ten" comes to mind. However, when we are speaking about organizational resilience, and national vulnerabilities, a more specific system seems appropriate.

With that aim in mind, I am going to suggest that large organizations and enterprises, infrastructure, government entities and the US as a whole, establish a common means of measuring the threat, by which then, response policy is developed.

In deference to the well known and finely tuned system of American defense we have used for decades, DefCon-I to DefCon – 5, I am proposing that we establish a parallel system to represent the Cyber-Health and virtual defense posture of companies and nations alike with a range of states-of-readiness, called perhaps:

CyCon - I

153

CyCon - II
CyCon - III
CyCon - IV
CyCon – V

Following the Time Based Security model, the goal is to integrate a detection and reaction system to include American business and the government. The following chart is a suggested example only, of the types of conditions we might expect to see to represent the Cyber-Health of both companies and the country.

CyCon-I represents the lowest level of detected offensive activity, and CyCon-V represents massive detected activities and dire consequences to the victim. Note that the scale is skewed one level. While a particular company or organization might be detecting and reacting to CyCon-II or CyCon-IV events, there is little effect on the National CyCon level. Regardless of how painful the attack might be for one company or its customers, staff, etc., there is a negligible effect at the national level. Thus the proposed scales for National CyCon Detection and Reaction require more organized attacks. Note here, that with the use of CyCon labeling, there is no need to resort to the word War.

What the CyCon model suggests is a more coordinated approach to organizational and national preparedness. Much of the work this author has done on the nature of Time Based Security as an alternative to the conventional military-driven computer security model of Fortress Mentality, reinforces the need for extensive sensors, or detection mechanisms throughout the enterprise, regardless of its nature.

	Corporate/Organizational	National
CyCon - I	Noise level; no attacks above chosen threshold	Occassional reports of CyCon-II from Corp. and/or No systematic assaults against government facilities or infrastructure.
CyCon - II	Unapproved scans, occassional hack attempts / Detection sensors triggered.	Some government facilities under hack-attack, limited Denial of Service and/or Reports of CyCon-III from Corporations. Infrastructure not affected
CyCon - III	Coordinated hacking and some Denial of Service attacks. Losses evident.	Noticeable successful attacks into government systems or medium-level DoS and/or more than one Corporate CyCon-IV reported. Infrastructure under some duress from assaults.
CyCon - IV	Company under extreme assault. Portions of networks isolated, customer services degraded, Denial of Service effective.	Government systems under coordinated assault, systems under heavy DoS, medium number of Corp CyCon-IV's or at least two Corp CyCon-V's or one major infrastructural attack causing severe degradation of service.
CyCon - V	Company under extreme assault. Must shut down all electronic facilities to isolate and preserve system.	Some government sites shut down/isolated by major attacks and DoS. Several to many Corp CyCon-V's. Major successful infrastructural attacks. National economy is affected

The sensors must be able to understand the nature of the attacks and behavioral anomalies throughout the virtual existence of the networks, infrastructures, and report back to a centralized repository and response station. This is accomplished at the enterprise network level by several popular products today. It is used, however, in only a very small percentage of the organizations that could really benefits from its use.

What is missing, in addition, is the means to create a centralized national reporting repository, whereby the national CyCon level can be measured on a real-time basis such as in the following figure.

CyCon Activity

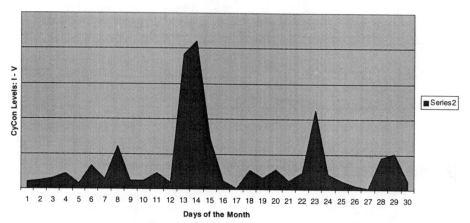

With a real-time monitoring and detection system and the proper reporting channels to a centralized facility (enterprise or national) broad CyCon levels can be established. With appropriate weighting for time, intensity, value, and other considerations, a company or the country can quickly look at the detected activity just as today we employ network-monitoring tools to gauge the real-time performance of a network, another TBS technique.

In more sophisticated applications, heuristics will come into play, the systems will be more self-adapting and learning, automatic remote responses will be monitored as well, and momentary spikes of high detected CyCon levels will be dealt with quickly and automatically. Thus, if a severe attack occurs against a major domestic firm, if its own Detection/Reaction systems are in order, the reports that it feeds to the national CyCon repository would barely register a blip.

On the chart above, we see a full month of activity. With a minor detection that triggers a CyCon-I alert on the 8th of the month, everything seems pretty normal until the 13th, 14th and 15th experience an extraordinary set of attacks, thus triggering a CyCon-4 condition. It appears that the attacks were dealt with in short order as the state of affairs went immediately back to nominal. That's good Time Based Security in action. All, without using the term War. The rest of this month was not totally without incident; on the 23rd. a serious set of events raised the CyCon level, and on the 28th and 29th, a similar, but not as serious set of attacks caused a very noticeable spike.

How an organization or nation responds or reacts to CyCon alert conditions is independent of TBS, but certainly a reporting mechanism which reflects both the detection and reaction process is a stronger defense. The TBS Model is based upon the concept that a strong defense comes from the ability to detect attacks in progress, not after they occur, and to react to those attacks in a timely manner.

Consider conventional law enforcement. Say you're in New York City, and you get mugged. At the police station, you tell the desk sergeant that the mugger held you ant knifepoint, got your watch, two credit cards and $200. What will he say? "Count your blessings. You're alive. Now get out of here."

Conventional law enforcement is based upon the premise of waiting for crime to occur and then picking and choosing those crimes that have some hope of being solved. To heck with the rest. And this is the same that we see with cyber-crimes: law enforcement is not capable of responding to every reported cyber-crime or assault against a company – or even the Pentagon. With such a "physical" and limited mind-set, law enforcement essentially says to the hacker and the potential criminal, "go ahead, have a ball. If your crime doesn't reach our artificial limbo-bar of $50,000 (for example, as in the case of one law enforcement organization), we don't have the time, resources or money to help you." My clients and I have both experienced the intense frustration of law enforcement out-gunned by teenagers with an attitude.

Now, at the national level, let's escalate the attacks and examine what happens if a serious set of attacks are detected, and perhaps triggers a CyCon-III or higher alert. The first question, is "who's in charge?" Is it the FBI as the recent Reno proposal suggests, with only minimal help from the Defense community? Well, maybe that's appropriate for much domestic cyber-crime, and even low-threshold cyber-terrorism.

But what happens when the escalated infrastructure attacks are from overseas? The FBI has had a couple of successful coordinated responses with the Russians, the Italians and Argentinians in well known cases. But these responses again echoed the weak laissez-faire-ism of "give us your best shot, and we'll try to track you down."

In the cyberworld, a strong defense is going to increasingly mean a strong offense and response. That is, if the parties in charge of defending our economic well being and infrastructures are limited to "catching the bad guys" if they can, where does deterrence come into play. Sure, we think of Mutual Assured Destruction as a Strangelovean relic – but it worked. We built a formidable and credible responsive capability in both conventional and strategic weaponry. As Martin Libicki has repeatedly said, "we can always make the rubble bounce."

So, in the cyberworld where is our credible deterrence?

Assuming that the CyCon alert system or something like it goes into effect, are we merely going to stand around and take it, or are we going to develop a National Information Policy based upon strength? Admittedly, we do not have all of the answers today. Unless we create a system of deterrence, however, all of the CyCon alert systems, and National Information Protection Centers and well meaning intentions of the FBI, the CIA and Defense and Intelligence Community will be for naught.

Say, we find ourselves at CyCon-III for some reason. We need not immediately resort to or respond with Information Warfare, or Cyberwar or Information Operations, regardless of who the lead US entity is. We can choose to initiate a "CyCon-III Response" which may include a series of both offensive and defensive maneuvers – assuming we have a policy of deterrence. Either way, we do not need to resort to war, or any longer enter the debate of whether we are at war or not, because it no longer matters.

We might find ourselves in a CyCon-II condition with a non-government organization that is physically located within the borders of a somewhat friendly nation-state. We are not at war; we are in a heightened set of defense and perhaps offense, as defined by the policy and the Option Charts. The biggest question is who responds and what is the response: Deterrence. Regardless, the CyCon alerting system gives us a way out of our linguistic conundrum regarding the word war and provides a workable defensive model at the enterprise, inter-enterprise and infrastructural levels.

Assume a CyCon-IV condition exists with a small nation-state, and their detected activity is edging towards CyCon-V. Policy should dictate our response without being required to Declare War. Back to policy, of course. The Reaction Matrix at the national or infrastructural level will look quite different than for an organization, but the principle and methodology is identical.

While we in the "Infowar Choir" well understand the use of virtual force in the offensive theater, we should also assist in making the options politically and diplomatically palatable to the decision and policy makers.

The use of CyCon levels to refer to defensive posture and engagement in a virtual world with intangible adversaries represents a reasonable way to avoid the use of the dreaded word War, yet still accomplish our goals.

28
Lies

"All warfare is based on deception."
Sun Tzu

*"In war (conflict), truth is so precious, it must be protected by
a bodyguard of lies."*
Winston Churchill

"Make a noise in the East and attack in the West."
Anonymous Chinese

I believe in lying. Sort of. Let me explain.

The bad guys will do anything they can to get you. You know that and it doesn't seem quite fair. They get to cheat, and you, as a network or systems administrator working for a real company have to play by the rules. They can lie. They can use verbal social engineering or hard copy social engineering or pull any sort of nasty trick they want to break into your networks or otherwise try to make your life miserable.

- Time Based Security suggests some innovative means to defend your networks, if we just apply some common sense.
- You goal is reduce the amount of time the bad guys have to attack you.
- You want your detection and reaction mechanisms to be as fast as possible.
- You may choose to invite the attacker to stay around for a longer period of time to give you more opportunity to collect forensic evidence and/or identify him.

All I'm saying is that you and me and we should create an even playing field. *"Do unto others as they do unto you,"* and in Cyberspace and Infowar, such logic makes impeccable defensive common sense. If the hackers lie to you why shouldn't you lie right back? There is a way. It is your right and defensive duty to

- Lie to your adversary.
- Deceive him in any way possible
- Force him to waste time/resources
- Makes his attacks a much riskier proposition
- Protect Your Assets by the same means he attacks you
- Use automatic responses and hands-off management
- Apply Time Based Security concepts
- Use "Deception"[1]

The Military View:
- The world is currently full of nations that are militarily weak, but ruled by despots who do not lack for cleverness or the willingness to use deception to maintain and expand their power.

Winn's Translations for Networks:
- *The Internet is currently full of networks that are defensively weak, but ruled by the technically and financially challenged who only need the willingness to use deception to maintain their systems integrity and expand their power*
- *The Internet is currently full of hackers, punks and goofballs that are morally handicapped, ethically weak, but who do not lack for cleverness or the willingness to use deception to maintain, project and expand their power.*

The main goal of your network defense is to keep your company functioning; keep the business process intact, and maintain day to day integrity so that there are no interruptions. Time Based Security gives us many of the tools to accomplish those very goals. Now, I'd like to explore another tool that can create victory without battle and impose your will on your network adversary.

---

[1] *"Deception" is trademarked by Interpact, Inc.*

That technique is deception. Lying. And if you think hard, and ask your legal counsel, there is no law against lying... especially to the bad guys.

Deception has been used throughout the history of warfare, from ancient times to today. Certainly the Trojan Horse fits the definition. Military leaders such as Phillip of Macedonia, Alexander the Great, Hannibal, Julius Caesar, and William of Hastings all the way to Saddam Hussein have successfully used deception to gain military advantage.

When undersized armies took on a larger force, they would have their horses pull logs behind them over dusty roads to give the impression that more manpower was coming to battle. Small armies would light 1,000's of fires at night to again give opposing forces the false impression of size. Psychological operations fit right into the deception mode with the philosophy, *"It doesn't hurt if your enemy thinks he's smarter or tougher than you."* Think about that. Playing it stupid is good?

During World War II, MGM Studios was making military support films, but worried about becoming a potential target for aerial bombing if the war came to America. Studio executives complained to the War Department who said they would take care of it. The government response was to send out experts who painted the roof of MGM studios to look exactly like a munitions plant. Joke?

D-Day planners convinced the Germans that the invasion would not be at Normandy, but some distance to the Northeast. When a German Enigma encoding machine was captured by the Allies, we figured out how to decode high level German transmissions. But we never let the Germans know that we could read their private mail, even if it meant sacrificing civilian targets to keep the secret. In modern warfare, electronic chaff is tossed from airplanes to confuse enemy radar. The Soviets poured thousands of electronic diodes into the concrete construction of the new American Embassy in Moscow. The intent was to confuse American counter-surveillance devices, which can't tell the difference between the non-linear junctions of the diodes and those in a real eavesdropping transmitter. Problem was they over did it, we found what they did out early in the construction process and we canned the new Embassy. Some experts maintain that Star Wars was nothing more than an elaborate technical public relations hoax of the first order to convince the Soviets we were willing to spend a gazillion dollars on space-based defense.

And then there was the Gulf War. Did the Patriot missile system work as well as was claimed? Probably not, but the media and folks at home ate it up. Saddam's grand deception scheme kept us shooting our smart bombs at Scud launchers that were nothing more than cardboard facades or shells of real ones. Deception clearly works.

Now, let's figure out how to apply deception to network security. It's time to even the odds! It is legally arguable to aggressively go after the bad guys. Corporate vigilantism is still only mentioned and knowingly approved by law enforcement in dark corners. They can't officially sanction the good guys to break the law to nab the bad guys, but the desire is certainly there. Nonetheless, an active defense is absolutely called for.

One of the common tools that the bad guys use to attack networks is scanning tools. Whether it's a purloined legal scanner from a real company or an underground tool, attackers want to understand and map out their victim's sites before "entering." So what happens? You spend hours and weeks to scan your own networks, fix as many vulnerabilities as you can... but there are always a few left. You can't remove all functionality... Time Based Security teaches us a lot about that, right? We don't want to resort to Fortress Mentality.

And what happens? After you've done your best, the bad guys come along, use their scanning tools and your defensive efforts now tell them exactly where to attack. They won't go after the things you have fixed; they'll go after the open electronic doors and windows... which their scanner points out to them. Your best Protective security efforts are now working against you! You've reduced your target suite and told them exactly where to attack. Counterproductive, don't you think? So try using some Deception against them! Some of the benefits are seemingly obvious.

- Works against insiders and outsiders
- Applies tried and true techniques
- Masks the leftover holes
- Gives ersatz appearances of network view
- Multiplies target suite
- Ambushes the attacker
- Makes attacks riskier propositions
- The enemy is never really certain
- No false positive or negatives
- Automatic hands-off management detection/response

And what is a secondary result? Scanners, legal or not, suddenly become useless. Deception includes entire suites to thwart scanning such as:

- Showing network vulnerabilities by the 100's
- Telnet open
- Default passwords are in effect
- Root is write enabled
- Most ports are open

Of course, it makes sense to reconfigure deception periodically so no one catches on to what you are doing. Or, on the other hand, you might choose to announce deception at logon to scare off would-be attackers. And as Time Based Security suggests, use deception mechanisms to keep attackers on-line for extended periods of time (reaction) to assist in identifying them. Deception is an excellent application of TBS:

- The detection mechanism can sit right on existing servers and respond automatically without any human intervention. D is very small.
- The reactions (deceptions) are automatic and very high speed. R is very small.
- The deceptions are based upon a Reaction Matrix which is policy driven.
- Contingency management and gaming exercises realistically will echo genuine attacks.
- Deceptions can increase the value of P, if Honey Pots are used. (See below.)

The Many Faces of Deception

Deception comes in many guises, and no one deception reaction is "Just Right" for everyone all of the time. (Common sense, by now, I hope!) ☺ Deception offers an entire suite of capabilities, and they should be picked judicially in any TBS-Deception application. The following is a Deception taxonomy that will be useful in its application. This taxonomy is based upon military experiences and history. As the TBS model and deception become more commonplace, I am sure that the following taxonomy will be greatly expanded.[2]

Concealment

Physical: Hiding through the use of natural cover, obstacles or great distance. Trees, branches; Terrain; Mountain Passes; Valleys

Virtual: Use best defensive practices for 'real' network services: Patches, Service Packs, Policy, Configuration. The object is to properly use and manage those basic security services that come with protective products and general applications.

Camouflage

Physical: Hiding movements and defensive postures (troops) behind natural camouflage.

Virtual: Hide the vulnerable points with network access rights, archiving, etc.

False/Planted Information

Physical: Letting opposition have the information you want them to have. Planting information you choose. False radio broadcasts, morphed pictures, videos and other misleading information aimed at enemies, leadership and general populations.

Virtual: Broadcast false network information from servers that are being scanned. Use the wrong IP address and the right IP

---

[2] *Please note that Deception is not designed to deal with Denial of Service.*

address and other conflicting information to confuse your network adversary.

Ruses

Physical: Where equipment and procedures are used to deceive the enemy; carry their flag/colors; march troops in the same formations; use the same uniforms and adversary radio frequencies (false orders). Initiate cries of help as if from the enemy troops.

Virtual: Tell the scanner that a legitimate scan is being conducted. Reinforce to the attacker that he is OK and safe doing what he is doing. Pretend to be another hacker working on the same system. Again, the TBS goal is to keep the hacker there for longer periods of time to gather forensic information.

Displays

Physical: Make the enemy see (or think he sees) what isn't really there. Horses pulling logs, 1000's of campfires, Fake artillery, Rubber Tanks, Dummy Airfields.

Virtual: Tell attacker you are calling the IP-Police; create a fake CERT alert; tell them you are 'Tracing' them; show fake firewalls and IP barriers

Demonstrations

Physical: Make a move that suggests imminent action; moving troops to the left, when you really want to attack on right; move troops constantly back and forth.

Virtual: Create an automatic defender, which seems to follow the attacker; create a daemon, which appears to launch a log/sniffer action or a trace.

Feints

Physical: An attacking demonstration. Use false attacks as a means of covering up the real mission/movements. Use false retreats to encourage chase by other side

Virtual: Appear to be only looking at the attacker when really switching defense modes. Appear to be helpless and defenseless when launching other means. Start an automatic response then stop and seem to try something else, but really maintain first one. Be loud about all moves by telling your adversary, or appearing to be so stupid, he thinks he's listening to your moves and you don't know it.

Lies
Physical: Lie to the enemy in any way that suits your needs. Use the media to lie. Use perception management. Creative perception management. Initiate protracted negotiations. Circulating false reports to the Net. Fabricating treasonable letters

Virtual: Use electronic lying in the same way. Let the system tell the attacker anything that furthers *your* goals. Creative perception management. Initiate protracted negotiations. Circulate false reports to the Net. Fabricate treasonable letters

Insight
Physical: Outthink one's adversaries. Study their past engagements and learn from their mistakes. Know your enemy better than he knows you. Stay one step ahead. It's a chess game: predict your opponent's moves.

Virtual: Understand their motivation. Learn the techniques. Collect logs of previous activities. Distinguish ankle-biters from serious attacks from professional attacks.

Honey Pots
Physical: Make something so attractive a target, your enemy comes running into your trap. Think Indians and sneak attack/ambushes.

Virtual: Clifford Stoll placed seemingly valuable national secrets on his computer to draw in the attackers. Create files with attractive information. *Come and get it!* Privacy Violations: medical, salary, etc. Rich, intellectual property. Corporate secrets. New products. Classified military information. Secrets of Saddam. Then Trap, Track & Trace.

167

Anything goes! Lies'R'Good in these cases, so use them. The construction of custom Deception suites is an attractive means for specific applications and industries that want to use Deception:

- Suck the attacker into a mirror of your Web banking applications to get the bad guys into a harmless area where you can watch, collect information and trace. The main Web banking application remains uncorrupted and functional.
- Brokerage firms can Honey-Pot the attackers into private information files/directories, which are really meaningless. Suck them in with "private confidential investment information." Maybe even encourage them to lose money!

Law enforcement, military and government sites can use the same approaches by picking and developing appropriate Deception suites that meet their specific goals. I would recommend that you speak with legal counsel with real Cyber-knowledge, about the proper means to collect forensic information that can be used in subsequent prosecutions.

Deception is deceptively simple to use as long as you understand some of the fundamental Rules of Deception.

1. Hide your moves from your opponent.
2. Never let your opponent see you as you are.
3. It's all about time. Waste their time in their useless attacks and keep them around so you can trace them.
4. One TBS reaction can be to announce your Deceptive existence to scare them away in short order.
5. Using TBS principles, Deception should operate at very high speeds, lowering (D + R).

Remember, *"There can never be enough deception."*
Sun Tzu

29
Final Thoughts

Employing TBS is not just a mere matter of applying a formula, and 10 seconds later having a calculator spit out an answer. The formulas and applications of Time Based Security are tools to be applied in systems analysis and practice. But the one area that will provide the greatest variability is P, and is unfortunately the one with the greatest number of unsubstantiated claims and erroneous assumptions.

Just how can we measure the *absolute* effectiveness of a security mechanism? We can't – the trial and tribulations of Fortress Mentality tell us that quite clearly. The provability of security effectiveness is terrifically complex. For example, is the effectiveness of cryptography solely measured by its key length? Or do we also need to consider its peer-review opinions, and the value of protection given to the keys? The Government has been trying for decades to develop absolute security in access control mechanisms. What value of protection do these systems provide? Over time, these protective values will change as (offensive) technology improves (and uses its own form of TBS implementation).

A firewall or other security mechanism is not a static defensive tool; it is subject to a highly dynamic set of conditions. P is ever-changing, generally for the worse where P $\Rightarrow$ 0. Faster computers, higher bandwidth, new attack tools, programming errors, insider information – these all contribute to the efficacy and strength of P.

We have to consider manufacturer claims like, "impenetrable", "hacker-proof", and "more secure than the others" which only contribute to the confusion. Most manufacturers take a "trust us" attitude, yet in the last few years, hundreds of vulnerabilities have been proven to exist against the very systems which were supposed to be protected. In this case, P does $\Rightarrow$ 0.

There is no perfect way to measure P - yet. So, a couple of different approaches to P are in order.

If you need to, set P as a reasonable range in TBS analysis. For example, the manufacturer of a firewall might claim that it has never been bypassed. In their mind, P = 1 week, or 1 month or some other high value. If you believe their claims, and you believe that all your systems configurations are 100% correct, and you believe that your network has never changed, and you know everything in the world there is to know about your network, then setting (D + R) = 1 week (1 month) might make sense on the surface. But the extended values of D and R place implicit and assumptive faith in P (Fortress Mentality) which is a blind static view of a network. (Not to mention insider attacks.)

Consider a more reasonable limited range of P; for example, P = 0 to P = 30 minutes (or 10 minutes or 10 seconds). Then perform the TBS analysis and see what occurs. Obviously,

D + R ⇒ 0 is the goal of TBS security implementation.

Worst case analysis, where P = 0 will provide the extreme (but reasonable) downside risk, and let the auditors do their job. Since P constantly changes, generally for the worse, this approach is the most conservative and places the onus of security on the (D + R) function; just as the TBS suggests.

Periodic security reviews to evaluate all three functions, P, D and R are a highly recommended manner by which to provide 'healthy check-ups' of especially the P component. As attacks advance in number and sophistication, and as P components are enhanced, double-checking the integrity of the P function provides additional confidence in the security of the systems.

But part of our job as security professionals is to help managers and organizational leaders make informed decisions as to the best ways to spend defensive security budget dollars. Historically, the argument has been to add more (protective) security mechanisms, add staffs to manage them, and then add more of each as attacks and related events spiral out of control. Time Based Security says that there are other, and perhaps better, questions to be asked and answered.

For example: should management invest $X in a security mechanism that might provide 30 seconds of additional protection? Or, should they invest $X in mechanisms that reduce detection (or reaction) by 30 seconds? What is the best 'bang for the buck?' How can you and management be quantifiably (measurably) sure of either choice? How can you make a sound judgment?

Or, what if the choice is between $X for 30 seconds more protection or $X-Y for 30 seconds faster detection/reaction? As products become attuned to the TBS way of thinking, security and systems administrators will have to make these sorts of decisions, and build their own charts and come to their own conclusions.

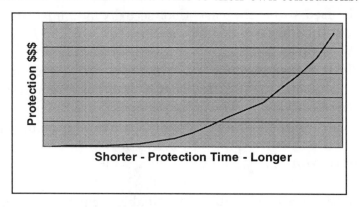

As the chart above shows, in general, adding protection time to a system costs more money, and in the upper right, there is a steep curve which says there is no amount of money that will provide infinite security. We all know that now.

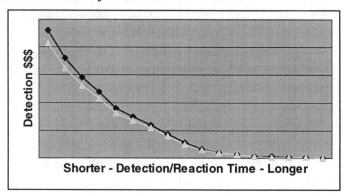

In the above chart, the exact opposite is true. Additional defensive capability is added by shortening the detection/reaction times, and those decreased times cost more. Similarly, there is no such thing as instantaneity; detection/reaction can approach but never reach zero-time.

The astute security and network administrator will utilize these charts and the TBS formulas to assist in making informed budgetary spending decisions. He will use the quantifiable metrics as a method to help convince management of the need for security, or increased spending in the aid of network or infrastructural defense.

It used to be the case that security administrators could brag about their achievements by saying, *"see, nothing happened, so I must be doing my job."*

Time Based Security will provide a very different picture. *"See, we caught 43 ankle-biting hackers and one renegade employee before they could do any damage to the firm."*

Which words would you prefer to use to your manager?

In Conclusion . . .

TBS offers measurable, quantifiable security as a function of time, which in a warp-speed world can determine success or failure. TBS offers the security professional, for the first time:

- A quantitative process methodology which can replicably measure the performance and effectiveness of security in enterprise, inter-enterprise and infrastructural environments, and
- A quantifiable framework to measure the cost-benefits of security expenditures, budgets and implementation.

The Time-Based Security Model offers an alternative to most of the conventional models that have been in use for the better part of two decades. While Orange Book models (i.e., Bell LaPadula) offered mathematical provability at the higher levels (such as B3 and A1, as did some of the higher levels within the

ITSEC model), such provability offered little to the real life security practitioner.

Instead of concentrating upon Protection as the first measurable level of security, it actually becomes t¹ ast component to be considered. Protection is needed, but now ι. is only one of a series of steps that must be taken in network and infrastructural defense. In a distributed open-environment, symmetrical communications infrastructure, detection and reaction become the primary metrics, with exposure-time the next critical quantity. Only then can we really talk about effective, cost justified protection meaningfully.

I appreciate all of the support that I have received in the last several years in these wonderfully arcane areas in which we live, breathe and work. Bob Ayers has left government employ and DISA, after almost 30 years of service. He is moving on to bigger and better things at Admiral Management Services in the United Kingdom, and I certainly wish him all the best in his new career.

Lastly, I would greatly appreciate any comments and thoughts you might have on how to apply TBS to meet your particular needs. Let me know about your successes, failures, insights, enhancement and improvements. Thank you.
WinnSchwartau@infowar.com

Winn Schwartau

About the Author (winnschwartau@nfowar.com)

Winn Schwartau is the President of Interpact, Inc. & The Security Experts, Inc. [www.securityexperts.com] (security consulting firms) and the COO of Infowar.Com. Inc. [www.infowar.com] He coined the term "Electronic Pearl Harbor" while testifying before Congress in June of 1991, and was the Project Lead of the Manhattan Cyber Project Information Warfare and Electronic Civil Defense Team. He is also a national radio talk host on Information Security for The CyberStation, New Media Broadcasting.

Affiliations
- Founder & Co-Sponsor: InfowarCon Conferences Brussels, London, and US. (1994 -)
- Editorial Columnist and Security Features Contributing Editor, Network World (1996 -)
- Publisher and Founder, Security Insider Report (1992 -)
- Security Columnist: PlanetIT, CMP Publications (1998 -)
- Member, Board of Directors, Tritheum Technologies, (1996 -)
- Editorial Board Advisor, Network Security (Elsevier), U.K. (1996-)
- Member, Board of Directors, HomeCom, Inc. (1996-1997)
- Member, Board of Advisors, IBIT, International Banking Information Technology, Liechtenstein (1995-1996)
- Member, New York Institute of Technology Criminal Justice Advisory Board (1997-)
- Member, Editorial Board of Advisors, InfoSecurity News. 1990-1997
- Technologist Advisor, Nat'l. Computer Security Association (1990-1997)
- Security Technologist to the International Security Systems Symposium Seminars. (1990-1993)
- Commentary Editor and Columnist: "Security Insider," Security Technology News, Phillips Pub. (1991-1994)
- Member, Editorial Board of Advisors, Crisis Magazine. (1988-1994)

Public Appearances
Television/Radio: CNN, ABC, NBC, CBS, BBC, CBC, Discovery, TLC, CNBC, MSNBC, Larry King, Encounters, Geraldo!

Speaking: US Congress, NATO, COMSEC, Banking Association, United bank of Switzerland, NASA, FBI, Sandia National Labs, Naval Postgraduate School, Swedish Government (DoD, Intel) IIR Australia, Dutch Police, Secure-Poland, Too Many Financial Organizations To Recall, US Air Force Academy, Electronic Funds Transfer Assoc., ISACA, Military Intelligence, Cooper's and Lybrand, Florida Law Enforcement, ASIS, IBM, ISSA, JWAC, Aerospace Industries Assoc., Society for Competitive Intelligence, RACF, Federal Law Enforcement Training Center, International Virus Bulletin, Open Sources Solutions, American Computer Telecommunications Association, Computer Security Institute, Federal Communications Conference, MIS Training Institute, ISSS, NCSA, Chambers of Commerce and hundreds of private organizations.